MATHEMATICAL KNOWLEDGE

Mark Steiner

Cornell University Press

ITHACA AND LONDON

First published 1975 by Cornell University Press.
Published in the United Kingdom by Cornell University Press Ltd., 2-4 Brook Street, London W1Y 1AA.

International Standard Book Number 0-8014-0894-6
Library of Congress Catalog Card Number 74-7639

Printed in the United States of America by Vail-Ballou Press, Inc.

To Rachel

Acknowledgments

It is a pleasure to acknowledge some intellectual debts. I should like first to thank Paul Benacerraf for letting me have the benefit of his keen critical sense throughout the preparation of this manuscript. Working under his guidance was a profound learning experience for me, in both how to think and how to write. How much I benefited is to be found in the difference between this work and earlier versions. Next, my thanks to Richard Grandy, whose influence is felt on every page of what follows. I have gained from Grandy's thoughts and from his pithy criticism of my own.

I have discussed the topics of my book with a number of philosophers, particularly Baruch Brody, John Bacon, Leslie Tharp, Saul Kripke, Michael Slote, Sidney Morgenbesser, Isaac Levi, Thomas Scanlon, James Higginbotham, and Charles Parsons. Credit has been given where their ideas have been used. Thanks, finally, are due Max Black and the mysterious logicist who read the manuscript for Cornell University Press and made valuable criticisms.

Much of my research was done in Jerusalem during the year

Acknowledgments

of my fellowship at the Hebrew University. I am grateful to Yehoshua Bar-Hillel for making this support possible, and to the Danforth Foundation, which supported me throughout my graduate years at Princeton.

This book contains extensive quotations from Ludwig Wittgenstein's *Remarks on the Foundations of Mathematics*, published by Basil Blackwell, 1956, and Chapter Four contains, in slightly revised form, my article, "Platonism and the Causal Theory of Knowledge," *Journal of Philosophy*, 70 (1973), 57–66. Permission to use this material is gratefully acknowledged.

MARK STEINER

New York City

Contents

9

Abbreviations

Bibliographical details are given in the References.

BP   P. Benacerraf and H. Putnam, eds. 1964. *Philosophy of Mathematics: Selected Readings.*

FLP   W. V. O. Quine. 1953. *From a Logical Point of View: Logico-Philosophical Essays.*

OR   W. V. O. Quine. 1969. *Ontological Relativity and Other Essays.*

PL   W. V. O. Quine. 1970. *Philosophy of Logic.*

STL   W. V. O. Quine. 1963. *Set Theory and Its Logic.*

WO   W. V. O. Quine. 1960. *Word and Object.*

WP   W. V. O. Quine. 1966. *The Ways of Paradox and Other Essays.*

Introduction

The aim of this book is to subject various beliefs concerning mathematics to epistemological critique. I regard as a datum the assumption that most people know some mathematical truths, and some people know many. This may seem hardly worth saying, but, as P. Benacerraf (1973) has shown us, it is often overlooked. One of the goals of what follows is to explore the consequences of such neglectfulness.

But in order to know, a person must *come to know* (Plato's theory of *mnemesis* aside). This, already the second banality in two paragraphs, signals equally a requirement for a philosophy of mathematics. If a theory of mathematics can be made to imply that no one can come to know any mathematical truths, the theory is false. Though the theorist may have ignored epistemological questions, his theory may nonetheless be committed to unforeseen absurdities.

The chapters that follow explore various (alleged) ways of coming to know mathematics—ways that have been suggested, implied, or, conversely, made impossible by the great philosophies of mathematics or by common opinion. I shall discuss, in varying detail, coming to know mathematical truths (a) via mathematical logic; (b) via consistency proofs for formal systems; (c) via proof in general; (d) via what is called "mathematical intuition." But let me explain.

Chapter One, the longest, deals with the epistemological consequences of logicism, with the effects on mathematical knowledge of basing it upon logic. Since I do not scruple to decide herein whether the logicist reduction of mathematics is really a reduction to set theory, and not logic, I dare to include W. V. O. Quine in the company of such unabashed logicists as Gottlob Frege, Bertrand Russell, and Alonzo Church. The major difference between Quine and the others is that the others view the reduction of mathematics to "logic" as revealing the true grounds of mathematical knowledge, while Quine views it only as showing how numbers may be "dispensed with," or displaced by sets, in the interests of progress.

Chapter One opens with a discussion of Henri Poincaré's criticism of logicism. One cannot hope, he says, to know mathematics solely on the basis of logic; there is a more fundamental principle, the law of mathematical induction, which cannot be reduced to logic. Epistemologically, the reduction cannot be carried off. There are, of course, many variations on Poincaré's theme; I choose Charles Parsons' as being the clearest. Parsons feels that Frege's and other logical systems introduce two concepts of the natural numbers: that expressed by the (defined) logical predicate

x is a natural number

but also the basic concept of the series

0, S0 (i.e., the successor of 0), SS0, SSS0, and so on.

To prove these two concepts coextensive, Parsons argues, metamathematical induction is required; irrelevant, then, that mathematical induction can be proved within the logistic system. Logic alone, therefore, cannot be the sole ground of arithmetic knowledge.

I argue contra Parsons in the body of this work that the logicist is not responsible for demonstrating *in his formal system* the coextensiveness of the two concepts of number, because mathematical knowledge is possible in logic without the knowledge of that coextensiveness.

Chapter One continues with a discussion of Ludwig Wittgenstein's epistemological attack upon logicism, a discussion aimed at proving that it is not possible to come to know mathematical truths through the medium of logic, because we cannot come to know enough logic. Here some extended explanations are necessary, to show why Wittgenstein's arguments must be taken seriously.

Classical logicism sees logic as the epistemic ground of all mathematics, including arithmetic (Hempel, 1945, *BP*, p. 366), and arithmetic as logical truth systematically abbreviated. But when asked for the ground of logical truth, logicism hides behind empiricism, assuring us that logical truths are "analytic." To know that all bachelors are unmarried, the logicist says, it suffices to know English; the same goes for logical truths.

The logicist knows that, on his own account, most mathematical truths are not recognizably analytic, even though recognizably true. To account for mathematical knowledge, the logicist stands with A. J. Ayer (1936):

The power of logic and mathematics to surprise us depends, like their usefulness, on the limitations of our reason. A being whose intellect was infinitely powerful would take no interest in logic and mathematics. For he would be able to see at a glance everything that his definitions implied, and, accordingly, could never learn anything from logical inference which he was not fully conscious of already. But our intellects are not of this order. It is only a minute proportion of the consequences of our definitions that we are able to detect at a glance. Even so simple a tautology as "$91 \times 79 = 7189$" is beyond the scope of our immediate apprehension. To assure ourselves that "7189" is synonymous with "91 × 79" we have to resort to calculation ... that is, a process by which we change the form of expressions without altering their significance. The multiplication tables are rules for carrying out this process in arithmetic, just as the laws of logic are rules for the tautological transformation of sentences expressed in logical symbolism or in ordinary language. As the process of calculation is carried out more or less mechanically, it is easy for us to make a slip and so unwittingly contradict ourselves. ... The more complex an analytic proposition is, the more chance it has of interesting and surprising us. [*BP*, p. 300]

In order to see the difficulty with this reply, consider a simple example:

$$1{,}000{,}000 + 1{,}000{,}000 = 2{,}000{,}000.$$

What does this truth amount to, viewed as a truth of logic? Logicians and other philosophers (Goodman and Quine, 1947) have often claimed that it is best seen as a transcription of a first-order schema, all of whose instances are logical truths ("tautologies"). The schema, in abbreviated form (see, e.g., Quine, 1959, pp. 231–232), looks like this:

$$(\exists_{1000000} x_{1000000}) F x_{1000000}$$
$$\&$$
$$(\exists_{1000000} x_{1000000}) G x_{1000000}$$
$$\&$$
$$-(\exists x)[F x \,\&\, G x]$$
$$.\supset.$$
$$(\exists_{2000000} x_{2000000})[F x_{2000000} \,\vee\, G x_{2000000}].$$

To write this out in primitive notation would take at least one thousand pages, to say nothing of the paper that would be wasted on a proof of it from the axioms of quantification theory! Obviously, one could not possibly recognize as such either the theorem or a proof of it. True, it can be proved to anyone's satisfaction, even without writing them out, that there must be a theorem and a proof in quantification theory corresponding to

Introduction

1,000,000 + 1,000,000 = 2,000,000.

But the metaproof of this fact assumes as a premise that one million plus one million equals two million—a fact we were supposed to learn from logic! Later I shall dig deeper to see whether the apparent circularity can be eliminated. But Wittgenstein's point has pertinency that cannot be dismissed.

Let me recapitulate the argument to make this clear: I began by emphasizing that the logicists regarded their reduction of mathematics to logic as having epistemological significance, in showing that "mathematics is based on logic," that logical grounds are sufficient to base mathematical knowledge on solid footing. In order to save this claim, Ayer—or a logicist—was forced to introduce the concept of "tautological transformation." Then I objected that to know that these transformations are possible, one must first, it seems, know the very arithmetic truths that are to be justified.

One might conclude that none of us humans knows any mathematical truths, since we are not able to follow the long logical transformations that alone justify mathematical knowledge. But this claim is surely too preposterous to be taken seriously; the whole discussion was begun with certain data in mind—among them, surely, that we all know that

1,000,000 + 1,000,000 = 2,000,000.

That is, *we* know, not some "being whose intellect was infinitely powerful," in Ayer's words (1936). This is the kind of fact that, though trivial, is often overlooked.

It is certain that logicists attributed epistemological significance to their "reduction." The reduction was supposed to provide a foundation for mathematical knowledge, to the extent that Frege felt that arithmetic was "tottering" when his logical system was proved inconsistent. Thus, too, Russell excoriates the mathematicians "who have learnt a technique without troubling to

17

Introduction

inquire into its meaning or justification" (Russell, 1919, p. 194). *Principia Mathematica* itself was supposed to supply such a justification, though Russell himself had begun with the observation that the propositions of simple arithmetic are more "obvious" than logic (1919, p. 2). But Russell was insufficiently concerned with the mechanics of such justification. He certainly did not reflect upon how one might come to know, for example, a proposition whose logical version might require more space to write it up than there are molecules in the universe. Had he done so, he might have seen that if the reduction is to justify mathematical knowledge in general it must justify every particular piece of knowledge. He might have seen that the form into which logicism casts arithmetic truths does not permit us to justify knowledge that we undoubtedly possess; for although the logicist reduction certainly suggests methods of coming-to-know in mathematics (the standard methods associated with analytic truths), these methods presuppose knowledge of what is to be justified.

I am not suggesting that logicism cannot withstand this kind of criticism—the kind associated with Wittgenstein. I am only suggesting that this criticism must be taken seriously.

The Appendix to Chapter One (printed at the end of the book) is an examination of the controversy between David Hilbert and Poincaré over how one justifies mathematical knowledge. Poincaré charges formalism no less than logicism with circularity. Mathematical induction for Poincaré cannot be squeezed out of logic, but neither can it be considered a Hilbertian "ideal sentence." Mathematical induction, after all, is unavoidable in proving Hilbert's system consistent.

Hilbert replies by distinguishing two types of induction: "contentual" (*inhaltlich*) and "noncontentual." It is the former only that is to be used in his projected consistency proof, and it is the latter that appears in the formal system as an ideal sentence. The burden of the Appendix is to sustain this distinction. Chapter

Introduction

One and its Appendix together thus provide refutations of Poincaré's briefs against both logicism and formalism.

Although the arguments of Wittgenstein and Poincaré ultimately fail—as I show in Chapter One—other arguments contra logicism do much better, and they are discussed in Chapter Two. One of the most telling arguments against epistemological logicism is that the theory to which arithmetic is reduced—whether we call it set theory or logic—is far less certain than arithmetic itself. If so, arithmetic cannot be based on logical foundations.

This argument, however, does not refute Quine's neologicism (though the arguments of Wittgenstein and Poincaré, if valid, would have), which pretends not to disclose the epistemic foundations of mathematics. Quine's only claim in favor of the reduction of mathematics to set theory is an Ockhamite lessening of the number of kinds of things that mathematics is forced to contemplate. The theory of classes has the power to unify mathematics, and this goal itself justifies the logicist program.

Quine thus feels that one can have ontological improvement even at the cost of epistemic weakening of arithmetic. In the first section of Chapter Two—after a tortuous dialectic—I come to the conclusion that this improvement is impossible. The final judgment is, thus, that logicism is untenable, in classical or Quinean form, even if the critiques of Wittgenstein and Poincaré are put to naught. The epistemological failings of logicism preclude not only basing arithmetic upon logic, but even dispensing with numbers in favor of sets.

The second section of this chapter, dealing with Benacerraf's well-known article, "What Numbers Could Not Be," treats of the philosophical theory that, according to Benacerraf, springs forth from the ashes of logicism. Here I determine that Benacerraf's arguments do not lead appropriately to the conclusion that numbers are not objects—indeed, that this conclusion is in flat contradiction to some of the premises of his argument. On the

Introduction

other hand, Benacerraf's prime contention, that the natural numbers have no properties but those they have in relation to one another, is sustained later, especially in Chapter Four.

Chapters Three and Four touch upon general issues in the epistemology of mathematics. Here the discussion proceeds without explicit reference to the classical philosophical schools. Chapter Three examines the relationship between mathematical knowledge and mathematical proof, to see whether the former requires the latter. I begin with a modest analysis of what it is to know a proof of a mathematical truth, and end by showing that knowing proofs is not necessary for knowing mathematics. A surprising corollary of the discussion is that the marshaling of inductive evidence, of the type relied upon in the natural sciences, is not only useful in mathematics but can actually confer certainty upon a mathematical truth.

Chapter Four treats directly of the acquisition of mathematical knowledge. The beginning of the chapter addresses itself to the anomaly I spoke of above—that knowledge of abstract objects is possible. This appears to be a contradiction. How can we acquire knowledge about entities that do not causally interact with the world? Is not all our knowledge a product of such interactions? Is not a Platonist account of mathematics automatically ruled out by epistemological considerations? The reader is referred to the text, where Platonism is rescued.

The rest of the chapter is devoted to reflections on "mathematical intuition," the alleged faculty of apprehending mathematical truth which Platonists and others have traditionally invoked to explain mathematical knowledge. After removing some a priori objections to the existence of such a faculty, and noting the number of impressive philosophers who accept mathematical intuition as fact, we are forced to admit that Platonist accounts of this supposed faculty have been sorely lacking. The chapter therefore concludes with a speculative discussion of the prospects for an *empirical* theory of mathematical intuition. Intuition emerges with fewer pretensions, but less shrouded in mystery. Mathematical intuition is not vindicated, but some of the evidence for its existence is marshaled. (I should like to point out that I did not have the benefit of studying Charles Chihara's book *Ontology and the Vicious-Circle Principle* [Cornell University Press, 1973] in preparing my own. In his book, Chihara considers from a different perspective many of the topics discussed here.)

Though the topics treated in this book are various, a single point of view lies behind all the discussions, one that the reader has a right to know now, the more so because I argue very little for it—directly—herein. This is the view that mathematics is a science, whose methods differ little, in principle, from those of other sciences. As I see it, mathematics studies the natural numbers as zoology studies animals (to revive a discarded Russellian position). Mathematics can be distinguished from the other sciences only by its subject matter—not on the grounds that it has none.

The reader can thus read the attack on logicism in Chapter Two, the weakening of the role of mathematical proof in Chapter Three, and the defense of Platonism in Chapter Four as flowing from, and reinforcing, a general conviction that the notion of mathematics as a peculiarly "formal" or "deductive" science is mistaken. And this is a conviction that I think Poincaré shares—so that, though this book begins with an attack on him, we conclude in partial agreement.

One

Logic and Mathematical Knowledge

There is a tradition critical of logicism which is strikingly epistemic. It attempts to undermine the epistemological significance of the reduction of mathematics to logic, without addressing itself directly to ontological questions. It is the line of attack associated with Poincaré, Wittgenstein, and many others, of whom Parsons (1965, pp. 180–203) is one of the more recent. All of these writers share the conviction that the reduction of mathematics to logic does not offer an explanation of mathematical knowledge which can withstand philosophical scrutiny.

Let us begin by characterizing logicism. Probably the most succinct expression of the logicist thesis is given by R. Carnap (1931, *BP*, p. 31): "1. The *concepts* of mathematics can be derived from logical concepts through explicit definitions. 2. The *theorems* of mathematics can be derived from logical axioms through purely logical deduction." Now the truth of logicism would surely have epistemological consequences. Given that logic is an a priori science and that mathematics "reduces" to logic, logicism could explain our mathematical knowledge, in particular how mathematics comes to be known a priori. C. G. Hempel (1945, *BP*, p. 366), for example, takes this line: "It is a basic principle of scientific inquiry that no proposition and no theory is to be accepted without adequate grounds. In empirical science ... the grounds for the acceptance of a theory consist in the agreement of predictions based on the theory with empirical evidence obtained either by experiment or by systematic observation. But what are the grounds which sanction the acceptance of mathematics?" Hempel's answer: the reduction of mathematics to logic provides the necessary "grounds." Propositions of logic are "analytic," hence known a priori, and it is possible to demonstrate the truths of mathematics as theorems of formal logic. (Note that both Carnap, quoted above, and Hempel, 1945, *BP*, p. 378, regard as part of logicism the view that the truths of mathematics are to be *theorems* of logic, *deduced* from the logical axioms using the rules of inference of logic.)

Logicism, then, is intended by its proponents to explain mathematical knowledge, as it clarifies mathematical concepts. Russell, in concluding his *Introduction to Mathematical Philosophy*, asserts boldly that "logic has become more mathematical and mathematics has become more logical. The consequence is that it has now become wholly impossible to draw a line between the two; in fact the two are one. . . . This view is resented by logicians who, having spent their time in the study of classical texts, are incapable of following a piece of symbolic reasoning, *and by mathematicians who have learnt a technique without troubling to inquire into its meaning or justification*" (p. 194, italics mine). Russell here claims that his work provides "justification" for mathematical knowledge, since the "identity" of mathematics with logic provides a foundation for the former. One might argue that this incautious burst on Russell's part contradicts his earlier claim: "The most obvious and easy things in mathematics are not those that come logically at the beginning; they are things that, from the point of

Logic and Mathematical Knowledge

view of logical deduction, come somewhere in the middle" (p. 2). This seems to admit that the Russell program adds nothing to the "obviousness" of grade school mathematics.

A more balanced view, which I feel is Russell's actual position, is this: we certainly had grounds for believing that $2 + 2 = 4$ before 1910 (when volume one of *Principia Mathematica* saw light). But we also had evidence for the differential calculus before Cauchy and Weierstrass—Newton is justly acclaimed. And our grounds for accepting the principles of analysis were not *truly* adequate until the foundations of that science were laid in the nineteenth century. Similarly, Russell might have argued, my work finally establishes the foundations of arithmetic, provides grounds for the a priori acceptance of mathematics. Just as there is no longer an excuse in analysis for avoiding the epsilon-delta technique of defining and calculating limits, so there is no longer any excuse for avoiding the logical calculus of Russell in justifying arithmetic.

In order for logicism to accomplish the epistemological goals of a Hempel or a Russell, logicism would have to consist at least of the following doctrines:

(1) There is some formal system of logic such that mathematics can effectively be generated from it.

(2) It is sufficient to understand proofs written in this system, in order to know all the truths of mathematics that we can know.

(3a) It is possible for *us*, with our limited abilities, actually to come to know mathematical truths in the way suggested by (2), that is, by constructing logical proofs of them.

And, if I am right about Russell, logicism might also be committed to:

(3b) One has not *true* mathematical knowledge of what one lacks the ability, at least latent, to produce a proof in logic.

I consider propositions (1) to (3) a *partial* reconstruction of the epistemological claim that mathematical knowledge is reducible to logical knowledge. They appear asserted with varying degrees of explicitness in the historical sources, and I maintain that logicists ought to be committed to them.

Proposition (1) reflects the logicist view that "the theorems of mathematics can be derived from logical axioms through purely logical deduction." True, one might regard logical knowledge as possible without formal deductions. Still, formal derivations were the stock in trade of almost all logicists, and certainly Russell's view that mathematics and logic are "identical" commits him to (1). (In light of Gödel's theorem, to save logicism, one might interpret "mathematics" in (1) as the classical axioms and results of mathematics. See the second section of Chapter Three for further discussion of Gödel's theorem.)

Proposition (2) expresses the logicist view that mathematical knowledge can in principle be derived solely from logical knowledge, without it being necessary to acquire any other kind of prior knowledge as evidence for the mathematical truths. Proposition (2) is crucial for a philosopher who hopes that logicism can explain the a priori character of mathematical knowledge.

Proposition (3a) differs from (2) in that (3a) states that we can—in practice, not only in principle—come to know mathematics by derivation from logical axioms. For one might grant (2) and admit that an infinite being could use *Principia Mathematica* in deriving mathematical truth, but that bounded beings cannot, and thus that logicism fails to explain *our* knowledge of mathematical truth. This is, in fact, precisely the claim of Wittgenstein, which I shall discuss later. Of course, anyone who attacks (2) is a fortiori attacking (3), since (3) implies (2).

Some philosophers might claim that in order to have facility with the object language itself, in order to produce convincing demonstrations in it, one must already have some mathematical knowledge. One must have, that is, expressed in a suitable meta-language, knowledge of the rules of inference and well-formedness. These are recursive rules, and mathematical induction is needed to be aware that however far one goes in the chain of inferences, or in building up well-formed sentences, truth or well-formedness is preserved.

But here we must be careful to distinguish between knowledge in the sense of ability and knowledge in the propositional sense— between "knowing how" and "knowing that." The claim of logicists is that someone could in principle come to know all of mathematics through his knowledge of the axioms plus his *ability* to form and recognize well-formed sentences, as well as his ability to execute and follow valid chains of inference. Though logic texts usually impart these abilities by making statements expressed, for example, in English, there is no reason at this stage to accept that as the only way to acquire such abilities. Nor does counting, in itself, and at its most primitive, seem to presuppose any mathematical knowledge (of the sort expressed in *Principia Mathematica*), though it certainly might produce such knowledge or help in producing it. So it is not yet a refutation of (2) to point out that, in order to be able to produce and understand proofs in a formal system, one must also have certain mathematical abilities, unless one can show that the abilities themselves pre-suppose certain factual knowledge. (P. Bernays made this mistake, as I read him; see his approving comments on Wittgenstein in *BP*, p. 523.)

Another note of caution needs to be sounded in regard to (3a) and (3b). One must specify clearly what one means by "we," "can," etc. For there surely is one sense of "can" in which "we" are not able, it seems, to produce all the required proofs, since all but the most trivial will be for us "unsurveyable." If the claim is that a being like us but with a longer span of survey-power could come to know proofs in the object language, it is unclear, at least at this point, what relevance this has to us, unless the logicist wants perversely to proclaim, on the basis of (3b), that we never really get to know most of mathematics, but that someone like us, only smarter, could. Are we, then, dealing with actual knowledge, or some idealized version? I shall return to the question of unsurveyability in my discussion of Wittgenstein.

Let us start, however, by considering that feature of the intellectual tradition under discussion which sees in mathematical induction an irreducible feature of mathematical knowing and which, therefore, rejects (2). Many versions of this objection to logicism exist, but the clearest is that of Parsons, who attributes it to Papert. Parsons (1965, p. 199) says:

> The Frege-Russell procedure defines *two* classes of natural numbers, such that mathematical induction is needed to show them identical. For we give explicit definitions in the set theory of '0', '$S0$', '$SS0$' . . . and *also* define the predicate '$NN(x)$'. How are we to be sure that '$NN(x)$' is true of what 0, $S0$, $SS0$, . . . are defined to be and only these? Well, we can prove '$NN(0)$' and '$NN(x) \supset NN(Sx)$'.
>
> If '$NN(S^{(n)}0)$' is the last line of a proof, then by substitution and *modus ponens*, we have a proof of '$NN(S^{(n+1)}0)$'. By induction, we have a proof of '$NN(S^{(n)}0)$' for every n.
>
> We can likewise prove by induction that every x for which '$NN(x)$' is true is denoted by a numeral. For 0 clearly is. If $n = S^{(m)}0$, then $Sn = S^{(m+1)}0$. So by induction (the derived rule of the set theory), if $NN(x)$, then $x = S^{(m)}0$ for some $m > 0$.

What is the objection here to logicism? This: in order to know mathematical truths, Parsons says, it is not enough merely to be able to produce and understand proofs in one of the various set theories. There is a proposition that *cannot* be demonstrated in the theory yet is essential to mathematical knowledge. The proposition can, however, be demonstrated with the help of induction in the metalanguage.

Now just what is this fundamental fact? Actually, a dual proposition. We shall consider the more important of these halves—that for every natural number n, the symbol formed by concatenating n occurrences of 'S' together with '0' denotes some member of the class denoted by 'NN', which is defined:

Logic and Mathematical Knowledge

$$NNx \equiv (F)(F0 \ \& \ (y)[Fy \supset FSy] \supset Fx).$$

(The converse, which Parsons also considers above, is less important, because no comparable tragedy would appear should the class of integers denoted by numerals be only a part of *NN*. What is important, if anything, is the neurotic fear that some numeral might not denote an integer in *NN*.)

But what Parsons does not explain is precisely *why* it is so essential to *know* this fact about the formal system. True, one who can justify his mathematical beliefs only by producing proofs in *Principia Mathematica* or in Zermelo-Fraenkel set theory (*ZF*) cannot even formulate the proposition as I did, in English, in the previous paragraph. Let us grant Parsons, then, that he could not know it. But this proposition does not appear in the premise of any mathematical proofs. What is lacking in a person's knowledge if (1) he knows the axioms of *Principia* or of *ZF* and (2) he knows how to prove all of mathematics in one of these systems?

It might be useful to note here a contrast to the Hilbert program. For Hilbert (1926) claims, like Parsons, that in order to *know* most of the propositions proved in the object language (those which make use of "infinitistic" methods) it is first necessary to know a metalinguistic proposition—that one cannot deduce a contradiction from the axioms. This, despite the fact that this proposition, of course, is not a premise of the theorems to be proved in the system and is not formally needed to prove them. (I ignore, for the present, the possibility that a proposition equivalent to consistency can be stated in the object language, for the simple reason that one without metalinguistic knowledge could not, it would appear, see the significance of this object-language proposition.) But the situation differs greatly because Hilbert claims that one cannot *know* that the axioms are true (in whatever sense a nonfinitary axiom can *be* true) unless one knows that they do not lead to contradiction. In our case, however, we have granted that our imaginary speaker of the object language knows all the axioms of the system. Given this, one asks whether facility with the Russell calculus is at least sufficient for mathematical knowledge—all of it—and Parsons claims that it is not. We want to know *why* not.

But, Parsons will say, suppose it were possible that one should construct a numeral that did not denote a member of NN. Suppose that the symbol composed of one million occurrences of 'S', followed by '0', did not belong to the set of names of members of NN? This would be grave indeed, but it cannot happen. Parsons himself gives the proof in the cited passage. Our mathematician, indeed, cannot produce the proof. But what still has to be shown is that this ignorance detracts from his knowledge acquired through the system. After all, for any *given* numeral '$SS \ldots S0$', the mathematician will be able to prove that the number denoted by it belongs to NN. As long as he knows *how* (note this distinction again) to do this, *we* know that *he* will not get into trouble. In other words, we have a metaproof that the definition of 'NN' in Russell or Zermelo captures our intent when we say "0, $S0$, $SS0$, *etc.*" We do not have to say that our mathematician-logicist, working in the object language only, has *two* concepts, as Parsons insinuates: that of a series of numbers, through a kind of inductive definition, and that of a set, NN, provided for in the theory. The mathematician works only with the concept 'NN' as defined and he need not worry about any other concept of the natural numbers that a kibitzer might have. If for some n, $S^{(n)}0$ were *not* in NN, the mathematician-logicist would not be working with the natural numbers, as we do. But since this does not happen, we may say that the logicist has been working indeed with the set of natural numbers as we know them.

It is not correct to say, with Parsons, "We give explicit definitions in the set theory of '0', '$S0$', '$SS0$', ..." We give explicit definitions only of '0' and 'S'. There is no concept in the set theory as such of the infinite set "composed of 0, $S0$, *etc.*"—the only

Logic and Mathematical Knowledge

concept is that of NN, which happens to contain the designata of all the numerals '0', '$S0$', ... But the concept of such a series remains outside the set theory, it is not a concept *of* the theory. Nor is the technique of inductive definitions primitive to the theory; it can be derived within the theory, since all such inductive definitions can be proved as theorems within *Principia*. The concept of the series denoted by the numerals, the concept "0, $S0$, *etc.*," is an informal one, then, replaced entirely in set theory by the explicit definition of 'NN'.

Futile it would be to deny, however, that in some way one must know the mathematical fact that the sequence 0, $S0$, $SS0$, etc., never leads out of the natural numbers NN, and not merely that if x is in NN, then so is its successor. This fact is certainly important to our concept of the natural numbers. But when put in *this* way (as a mathematical fact and not as a metalinguistic principle) it is actually provable in set theory. Instead of using the symbol $S^{(n)}0$, which is metalinguistic, we can define $S^{|n}0$ as the nth iteration of the *operation* S. This concept can be defined. For $S^{|n}0$ is nothing but x such that there exists a finite sequence u such that

$$u_0 \text{ is } 0$$
$$u_n \text{ is } x.$$
For every natural number $k < n$, we have
$$u_{k+1} = Su_k.$$

(All of this can be formalized in set theory.) Now it is relatively easy to prove the following theorems in set theory:

$$S(S^{|n}0) = S^{|n+1}0$$
$$S^{|0}0 = 0.$$

From this, plus the (provable) fact that the successor of a member of NN is a member of NN, it follows by *object-language* induction—the derived rule of the set theory—that for any n in NN, $S^{|n}0$ is in NN. (For details, see Quine, *STL*, sections 11–16.)

It is also provable that for every x in NN there is some n in NN such that x is $S^{|n}0$.

The situation, then, is as follows: the logicist-cum-mathematician is able to construct what we, looking from the outside, can prove to be names of all the members of NN, through a process that produces only such names—though the mathematician cannot articulate this, since he does not speak the metalanguage. Nevertheless, he can construct—if given a specific number denoted by some numeral '$SS \ldots S0$'—a proof that it is a member of NN. He knows that for each x in NN, Sx is in NN. True, he is open to "surprise" in the person of some numeral he constructs which names no member of NN, since he does not know the fact stated in the previous sentence but one. But I have argued that this (impossible) eventuality is not significant. First, and foremost, because it is, in fact, impossible—though it is arbitrary to further require that the mathematician himself know that it is, knowledge that is superfluous in acquiring knowledge within the system.

Further, his ignorance that his ability to generate names of members of NN does not flag—this ignorance does not bespeak a gap in his concept of number. For in the set theory there is no such concept of a sequence as "0, $S0$, $SS0$, *etc.*" We define only '0' and 'S', and from this, the set 'NN' denotes. It is *this* concept that is relevant, and this he has.

Finally, I argued, if we take the logicist-realist point of view seriously and note that we can define—in the theory—"the nth iteration of S," we can show—in the theory—that for however many iterations of S, we never leave NN. This *is* an important part of the concept of number, and it is this that Parsons actually appeals to in stating that logicism cannot account for the knowledge we must have that for n any number $S^{(n)}0$ is in NN. He appeals to the iterative property of the natural numbers, the property of closure, as a fundamental mark of the concept of number. But his conclusion is too hasty, for this property can be stated and proved of the numbers in the system itself: for any n, the nth iteration of S applied to 0 is in *NN*. I say, "Take the realist point of view seriously," because when we do, we stop worrying about the mathematician's inability to express the fact that the numerals never run out. As long as he knows that the *numbers* behave themselves, who cares what happens to the numerals? We never have use for more than a few hundred billion of them anyhow; there are other ways to name numbers than by numerals.

Frege's critic can still complain that, though we can show $S^{|n}0$ to be in *NN* for all n in *NN*, we still cannot show that, for n a number given by the numeral '$S^{(n)}0$', the conclusion follows. In this form, the theorem on iterations does not work, and an infinite regress is in the offing. But it seems to me that the mathematician-logicist has the right to define iterations in terms of any concept of natural numbers he wishes as long as the concept is workable. If he wishes to use the concept '*NN*' as defined in set theory in defining the general notion of iteration, he may. For this concept, judged by its deductive output, captures perfectly the informal concept.

We must now turn to a general objection that Parsons (1965, p. 200) has up his sleeve to our whole procedure. He represents the logicist as "replying" to his line of criticism by "denying that we have an independent understanding of inductive definitions. That is, if we ask Papert what he means by '"0", "$S0$", "$SS0$", etc.', he would have to reply by an explicit definition which would turn out to be equivalent to '*Num* (x)'." Parsons defines '*Num* (x)' to be (in my notation).

$$(F)(F`0' \ \& \ [Fy \supset_y F(`S'\frown y)] \supset Fx)$$

on page 199 for purposes we shall not mention here, but as what follows clearly shows, the argument and Parsons' reply could as well turn on '*NN* (x)' as on '*Num* (x)'. Parsons continues:

But this last reply to Papert depends on the claim that the apparatus in terms of which such explicit definitions as that of '*NN*(x)' are given can be

Mathematical Knowledge

understood independently of even the most elementary inductive definitions. This is implausible since the explicit definitions involve quantification over all concepts. It is hard to see what a concept is, or what the totality of concepts might be, without something like the inductive generation of linguistic expressions which (on Frege's view) refer to them.

There are two separate points hinted at here. The first is that the notion of a concept is incomprehensible to one who has not yet mastered the notion of an inductive sequence of linguistic expressions. The other is that the notion of the totality of concepts is incomprehensible to such a one. Insofar as this latter position is stronger than the former, it says that the totality of concepts is a notion that is mastered by reflecting on the generation of terms denoting complicated concepts from terms denoting more simple concepts with the aid of logical particles. If this is what Parsons is saying, then surely it is wrong. This would ignore completely that the definition of 'NN' is *impredicative*. That is, the quantifier in

$$(F)(F0 \ \& \ (y)[Fy \supset FSy] \supset Fx)$$

ranges over NN as well. This is no mere philosophical quibble: as Russell found out, to his sorrow, one cannot prove many things we need in elementary number theory without explicitly assuming that NN is a value of the variable 'F' in the above definition, for we must sometimes *substitute* 'NN' for 'F'. (Russell thought he had found a way out; see the Appendix to the second edition of *Principia*, but also see Gödel's criticism [1944] of this proof, in *BP*, p. 226.) It is precisely this impredicativity that renders it impossible to conceive of the aggregate of the values of 'F'—"the totality of concepts"—as being generated in a recursive style. Indeed, this is precisely Poincaré's criticism of the definition! (See *Science and Method*, pp. 190–191 ff.) As I have pointed out previously, it is Poincaré's criticism of the Fregean definition of 'NN' which embodies both Poincaré's aversion to impredicativity *and* his insistence upon the primacy of induction. Once one accepts the legitimacy of impredicative definitions, as Parsons apparently does, one cannot then turn around and say that the concept expressed in the definition is based in some upon "something like the inductive generation of linguistic expressions."

Charity, therefore, calls for a more modest interpretation. Parsons may mean only that the idea of a concept is closely tied to the notions of language, that one could not make clear what one meant by a concept in general to one who has not certain concepts about language. Without the notion of a concept in general, one could not grasp the definition of 'NN' which utilizes the device of quantification over all concepts. And on the other hand, one could not understand the various concepts relevant to language without first having the notion of an inductive sequence that seems to lie at the heart of understanding language—and one of these relevant concepts is that of a concept! No suggestion then by Parsons that the totality of concepts themselves must be seen as inductively generated.

Still, one has doubts about this speculative claim. These "concepts," after all, are really sets (concepts in extension) if we ignore Frege's peculiar views concerning concepts. Is it really true that one must have the notion of an inductive sequence of linguistic items before understanding what a set is? It seems to me rather that the notion of a set can precede the realization that language unfolds inductively. And, if we really do have the notion of the aggregate of all concepts, it does not seem that it is necessary to have this realization as well. (One might deny that we do, as Poincaré did, but Parsons it seems does not.) One need have only the notion of "concept" or "set," in my opinion, together with a grasp of the logical operator "all." For there is no good reason to believe that "all" changes its meaning from context to context. We have already pointed out that since the definition of 'NN' is impredicative, no idea of inductive "unfolding" of anything (concepts or linguistic entities) will *suffice* to convey the meaning of quantification over all concepts; now I suggest that it is not necessary either.

It seems that the persuasive force of Parson's claim is eliminated if one considers that the *ability* to recognize and to work with inductive sequences of linguistic items is not yet to have any knowledge about such sequences. It seems theoretically possible that one might have the concept, say, of a *predicate* without having the knowledge that his language unfolds in a manner analogous to the way the natural numbers unfold (after all, the analogy does not strike everyone), but he would probably have to have the ability, the "knowledge how" to construct sequences of linguistic items of arbitrary length.

I think it pointless to haggle over this. Let us, for the sake of argument, concede to Parsons that one could not see even what a concept is without "something like the inductive generation of linguistic expressions." But having made the concession, we may still claim a sense in which one does not truly understand inductive definitions, or the "etc." in "0, 1, 2, . . . etc." Once one has seen what a concept is, once one has climbed the ladder (in the Wittgenstein simile), and once one has used the notion of a concept to formulate an explicit definition of what used to be merely inductive—*once* this has all been done, it may turn out that one only now truly understands, say, what it is to be a natural number. This is merely the Quine-Neurath point that we use our bad concepts to make good ones. We pull ourselves up by our own bootstraps. Once we have the good concepts, however, they remain independent; we know how to use them in their own right. Understanding admits of degrees.

Both Parsons and his imaginary objector, then, set up a false antithesis: each of them champions a different concept of the natural numbers. But both are right in a sense—Parsons might be right in claiming that we need to have the idea of an inductive sequence in order to understand the definition of '*NN*'; the objector might be right in saying that we only understand the idea of inductive sequences when we have the definition of '*NN*'.

Consider again the case of continuous functions, limits, and so on. The early moderns certainly knew what a limit was, what a continuous function was, and they produced the calculus and some sterling results (perusal of a seventeenth-century text in mathematics leads to amazement over how much they knew). But in another sense, equally valid, their concepts were quite vague and heavily tied to "pencil and paper" intuition. The reconstruction of analysis in the Romantic Age could not have been done, in fact, without an intuitive grasp of the relevant concepts. Yet given the reconstruction, we are prepared to say seriously that Euler did not know what he was talking about as soon as he generalized beyond a certain level. (His individual results, however, are certainly his, as we shall have occasion to emphasize strongly in another connection.) We have tightened our standards for what is considered "understanding" to the point that no one is today considered to know a theorem of analysis unless he makes it clear that he could write it up in Weierstrass. This discussion, by the way, takes for granted that both the Fregean definition of '*NN*' and the Weierstrass definition of 'continuous' are adequate substitutes for what they replace.

It also bears repeating that the "inductive definition" of the natural numbers (and other inductive definitions) do not appear as primitive techniques in the formal theory, whether or not we say that we can understand them independently. We *prove* these "definitions" as consequences of the axioms. For example, we can easily demonstrate the inductive definition of 'natural number' from the definition of '*NN*'. '*NN*' is cleverly chosen to replace the pretheoretical inductive notion. In order to work with the theory, however, one need master only the meaning of '*NN*'. Nor need one know to what this concept corresponds, in order to use the definition of '*NN*' to yield theoretical knowledge. (This sentence is compatible with the possibility that one might need to acquire the informal concept as a stepping-stone to acquiring the concept 'NN', which I have relinquished to Parsons.)

We consider now the claim that the definition of 'NN', that is

$$NNx \equiv (F)(F0 \ \& \ (y)[Fy \supset FSy] \supset Fx),$$

involves "quantification over all concepts," and hence "the totality of concepts," as Parsons says. This claim is in fact true. One might quibble that the definition is of a form that allows us to restrict the quantifier to range only over "inductive" sets—sets that contain 0 and that contain the successor of every member, so that if we let script letters range only over inductive sets, the definition could be rewritten

$$NNx \equiv (f)fx.$$

This point, however, is not too important, since the set of inductive sets (intuitively, this comes out, the set of all extensions of NN) is quite big enough, if it exists at all as a completed totality.

There is a more important objection to Parsons' claim to be taken care of: Why do we assume that, because the values of a variable exist, they form together a totality? We will see in particular that the use of universal quantification does not necessarily presuppose reference to a completed totality of items in the range of the bound variable (see Appendix). But the definition of 'NN' is not (as far as I have been able to check) used in this way. Rather, UI is applied to the definiens in order to get specific instances. For example, in the proof of each instance of the induction schema (for the open sentence, say, 'Fx') one must generate the set $\{x : Fx\}$ and instantiate this set's name in the definiens. In short, the definiens could be written without the bound variable 'F', for all that we make use of it in negative subformulas in proofs.

Nevertheless, if we take into account that the definition of 'NN' is impredicative, we can perhaps make more plausible the view that the definition presupposes an infinite completed totality of concepts (sets). We must, however, assume the premise that whatever is a value of a bound variable must itself be complete. If we do, then since one of the values of the 'F' is NN, it follows that NN is complete. But for NN to be a totality, it seems plausible that the aggregate of values of the bound variable as a whole must be a completed totality, for it is these values that "generate" the set NN by infinite intersection. How to justify the main premise, I do not know.

To round off this discussion of Parsons, consider another Poincaréan argument that he develops, as he says, independently of the first one. Suppose one has proved by induction a proposition '$(x)Fx$', and by successive applications of *modus ponens*, demonstrated first from '$F0$' and '$(n)(Fn \supset FSn)$' that $FS0$, and then that $FSS0$, and so on until the desired result is reached. Parsons remarks (1965, pp. 200–201):

> We say that the proof by way of induction gives us the assurance that we *can* construct such a proof by successive applications of *modus ponens*. And the complexity of such proofs is unbounded. (But how do we see that we *always* can? By induction!) The application of the procedure involves the iteration of certain steps n times for the proof of $F(S^{(n)}0)$. Neither the reduction of induction to an explicit definition nor the Wittgensteinian doctrine that '$F0 . (n)[Fn \supset F(Sn)]$' constitutes the *criterion* for the truth of 'for all natural numbers n, Fn' gives us an assurance that there will be no *conflict* between the two methods for proving individual cases. When Poincaré said that a step of induction contained an infinity of syllogisms, he was saying that it guaranteed the possibility of all the proofs of the instances by reiterated *modus ponens*. He was right at least to this extent: if we do have an *a priori* assurance that there will be no conflict between construction in individual cases and inference from general propositions proved by induction, then this assurance is not founded on logic or set theory.

But the crux is *why* this lack of assurance should bother the logicist. According to my description of his position, of course, the imaginary mathematician could not even formulate what is said to lack assurance. Again we ignore possible transcriptions into the object language via Gödel numbering since Parsons would undoubtedly reply that to see that these express the appropriate propositions one would still need induction outside the system. We must first see what Parsons means when he says that we must know that there will be no *conflict* between the two methods of proving individual cases in number theory.

One thing he might mean is that there is a danger that one will somehow be able to prove, say, 'Fx' by induction, while by some other method '$-Fx$' is provable. But this is just the problem of consistency for a particular axiom. And this problem is ruled out in advance by our requirement that the mathematician *know* the axioms. Remember that the consistency problem arises for Hilbert precisely because he is *not* sure of the axioms, a different problem entirely. Because of such nagging doubts, and even doubts about *meaningfulness* of some of the axioms (when we attempt to give them an "intended meaning"), Hilbert requires some metamathematical results to bolster one's confidence. Given knowledge of the axioms, however, the logicist need not require any metalinguistic knowledge at all—or so I have been arguing.

Suppose, however, that Parsons is speaking not of inconsistency but of not being able to go on with the successive *modus ponens* inferences to prove in this way $F(S^{(n)}0)$ for some n, though we can plug in '$S^{(n)}0$' for 'x' in '$(x)F(x)$' proved by induction. But why should this be considered a "conflict," in Parsons' words, between the two methods? This would only show that one of the methods contains an incompleteness. There still would remain one way at least of proving all the individual cases; why ask for two?

The conclusion of this discussion should be clear: I have found no reason to accept Parsons' version of Poincaré as providing convincing arguments to reject logicism, in particular thesis (2) as logicism is formulated by me. The following section will conclude the same for Wittgenstein's arguments against the logicist's (3).

II

We turn now to a different criticism of the logicist philosophy, that of Wittgenstein. Wittgenstein's line is to deny, not (2) as Poincaré, but (3a) and hence (3b). Most often, that is, Wittgenstein does not deny that *if* all proofs in Russell's calculus were "survey-able" (if we could take all such proofs in at a glance), *then* arithmetic might be based upon logic. In practice, however, not all such proofs are surveyable; hence, Wittgenstein claims, we must intro-duce extralogical knowledge into the calculus, in order to use it (for example, knowledge expressed in the decimal notation). Although we call this process mere abbreviation, avoidable "in principle," we cannot truly eliminate from a proof all defined expressions and still preserve its property of being a proof (since, for Wittgenstein, a proof is an object that actually can convince). Finally, though one might prove mathematically that for each shortened proof a longer one exists (presumably by induction, though Wittgenstein does not say openly; this, by the way, is the connection between him and Poincaré), we could not say any more that the mathematical knowledge is based on the Russell proof. Hence, both (3a) and (3b) are false.

Wittgenstein also considers the objection that surveyability of an entire proof is not necessary in order for the proof to be grounds for mathematical knowledge. All that is necessary is that each individual step could be seen to be valid. Here he has a cluster of responses, which can, however, be distinguished. For one thing, he totes out that old philosophical bugaboo, the possibility that the Russellian proof alters its shape when not being observed, and part of an unsurveyable proof must always go unobserved. To scotch this possibility, we need more than logical knowledge. Second, even an individual step in a proof might be unsurveyable.

Third, even granted that one could follow all the steps, one would not know what one had proved. This, because (a) the last line of the proof, the theorem itself, might be unsurveyable or (b), a much more radical position, because one does not understand what one has proved, unless one understands the proof as a whole. Following all the steps of a proof does not mean understanding the proof, as anyone who has listened to mathematical lectures will agree. Besides, logic alone could not show the formal result equivalent to the (nonformal) mathematical statement. Wittgenstein even toys with the fantasy that what is proved in mathematical texts *cannot* be equivalent to what is proved in *Principia Mathematica*, since the criterion for the identity of what is proved depends upon the criterion for the identity of the proof—there cannot be two proofs of the same proposition. The proof in the textbook differs from its alleged double in *Principia* owing to the unsurveyability of the latter. (Actually, much of Wittgenstein's discussion of proofs is marred by his extreme nominalism, which leads him to speak of a proof as a token, a material object, whereas his remarks on criteria for identity are often incompatible with that point of view.)

We cannot treat all that Wittgenstein says in Part II of his published *Remarks on the Foundations of Mathematics* about logicism, and the hasty synopsis I have given hardly does justice to the wealth of true and false ideas in his writing. Instead we shall examine a few of Wittgenstein's actual statements in support of the positions sketched above.

The first issue is the extent of the unsurveyability of the proofs in *Principia Mathematica* of propositions whose proofs are surveyable in ordinary texts. Wittgenstein takes unsurveyability to be an inevitable result of replacing "ordinary arithmetic" by "Russellian arithmetic" (II, 13). By ordinary arithmetic he means decimal arithmetic, which can be introduced into *Principia*, according to him, only through a process of abbreviation. In a typical passage, Wittgenstein argues as follows: "Now let us imagine the cardinal numbers explained as $1, 1 + 1, (1 + 1) + 1, ((1 + 1) + 1) + 1,$

and so on. You say that the definitions introducing the figures of the decimal system are a mere matter of convenience; the calculation 703000 × 40000101 could be done in that wearisome notation too. . . . Now I ask: could we also find out the truth of the proposition 7034174 + 6594321 = 13628495 by means of a proof carried out in the first notation? . . . The answer is: no" (II, 3).

The decimal calculation is surveyable, but the Russellian is not; if we translated the decimal proof into its Russellian equivalent no human being could grasp the proof. And the reason for *this* is that the theorem 7034174 + 6594321 = 13628495 in primitive notation would have to be written $S \ldots S0 + S \ldots S0 = S \ldots S0$. (Imagine this written out.)

But does it? Note, first, that the decimal system in itself is not a method of calculation, though it certainly lends itself to a certain technique of calculation. The decimal system, like the stroke notation or the Principian notation, is simply a method for generating numerals. One can grasp the method without knowing yet how to calculate. The advantage of the decimal system, and of every system that has a base, is that each numeral has a dual meaning, provably equivalent. On the one hand, each nonzero numeral designates the successor of some other number, whose numeral can be obtained in a simple, mechanical way. Thus the numeral '11' denotes the successor of 10 and corresponds to '$SSSSSSSSSS0$' in set theory (where 'S' may denote any of an infinite variety of functions satisfying some simple conditions). But on the other hand each numeral is also a *polynomial in* '10', e.g., '533' is short for '$5 \times 10^2 + 3 \times 10 + 3$'.

We can switch back and forth from one meaning of the decimal numerals to another as the occasion arises. We can prove that each number is representable by a polynomial of '10' and that each polynomial of '10' yields a number. Once we have polynomial representations of numbers, we can use many shortcuts based upon the distributive, commutative, and associative laws. In decimal arithmetic one extracts one polynomial from another.

Mathematical Knowledge

One breaks up the addition of numbers into the addition of single columns of numbers; multiplications into the addition of partial products, and so on.

When we take this point of view, the fact that 4000, i.e., 4×10^3, is the successor of 3,999, i.e., $3 \times 10^3 + 9 \times 10^2 + 9 \times 10 + 9$, is not a definition, but an instance of a theorem about commutative rings, if you will. For our purposes we must keep separate the aspect of being a numeral from the aspect of being a polynomial, even though in general it is convenient to ignore the difference. Under one aspect, the numeral '23,444,444,444' is no more "surveyable" than the numeral formed in Russell's system by concatenating 23,444,444,444 times 'S' together with '0'. Assuming ignorance of the basic properties of the number system, we would be forced to do sums and products by direct application of recursive definitions, e.g.,

$$56 + 62 = (56 + 61) + 1 = (((56 + 60) + 1) + 1) = \ldots$$

precisely as we do in "Russell's system." And on the other hand, assuming knowledge of the behavior of polynomials over N, we can, even in Russell's system, perform the desired calculations with large numbers. All we need is to define '0' as '0', '1' as '$S0$', '2' as '$SS0$', ..., '10' as '$SSSSSSSSSS0$'. This done, we can accept Wittgenstein's challenge to prove

$$7034174 + 6594321 = 13628495$$

in *Principia*:

$$S^70 \times S^{10}0^{S^60} + 0 \times S^{10}0^{S^50} + S^30 \times S^{10}0^{S^40}$$
$$+ S^40 \times S^{10}0^{S^30} + S0 \times S^{10}0^{S^20} + S^70 \times S^{10}0^{S0} + S^40$$

$$S^60 \times S^{10}0^{S^60} + S^50 \times S^{10}0^{S^50} + S^90 \times S^{10}0^{S^40}$$
$$+ S^40 \times S^{10}0^{S^30} + S^30 \times S^{10}0^{S^20} + S^20 \times S^{10}0^{S0} + S0$$

$$S0 \times S^{10}0^{S^70} + S^30 \times S^{10}0^{S^60} + S^60 \times S^{10}0^{S^50}$$
$$+ S^20 \times S^{10}0^{S^40} + S^80 \times S^{10}0^{S^30}$$
$$+ S^40 \times S^{10}0^{S^20} + S^90 \times S^{10}0^{S0} + S^50.$$

Logic and Mathematical Knowledge

The reader will agree that I could write out all signs of the form '$S^{(n)}0$', since n is never greater than 10 (we do not need to learn anything after the 9's multiplication table or at most the 10's); the only reason I do not is compassion for the printer. Even if we replaced such signs, the reader will note, the result would be surveyable since the primitive symbols would form a pattern.

An objection leaps to mind. The multiplication sign is not part of the primitive notation, nor is the superscript notation for exponentiation. If we wrote out, let us say, $S^70 \times S^{10}0^{S^60}$, using the sign '+' only, we would need 7×10^5 occurrences of '$S^{10}0$', connected by $7 \times 10^5 - 1$ plus signs. Surveyability aside, the very notation for exponentiation seems to introduce a concept foreign to the primitive concepts of the logistic. As Wittgenstein himself puts it (1956, II, 47):

> Tell me: have I discovered a new kind of calculation if, having once learnt to multiply, I am struck by multiplications with all the factors the same, as a special branch of these calculations, and so I introduce the notation 'a^n = ...'?
>
> Obviously the mere 'shortened', or *different*, notation—'16^2' instead of '16×16'—does not amount to that. What is important is that we now merely *count* the factors.
>
> Is '16^{15}' merely another notation for '$16 \times 16 \times 16 \times 16 \times 16 \times 16 \times 16 \times 16 \times 16 \times 16 \times 16 \times 16 \times 16 \times 16 \times 16$'?
>
> The proof that 16^{15} = ... does not simply consist in my multiplying 16 by itself fifteen times and getting this result—the proof must shew that I take the number as a factor 15 times. ...
>
> So I am surely setting up a new connection!—A connection—between what objects? Between the technique of counting factors and the technique of multiplying.

Were Wittgenstein right, I would be wrong in saying that even in *Principia* ordinary arithmetic is surveyable and faithfully copies ordinary calculation. For even if it were the case (which it is not) that the replacement of exponentiation by multiplication and multiplication by addition preserved surveyability, the very use of exponentiation introduces a new concept into arithmetic that cannot be mirrored in the Russell calculus.

There is something right (though the conclusion, we shall see, is false) in what Wittgenstein is trying to say (II, 47).

When I ask "What is new about the 'new kind of calculation'—exponentiation"—that is difficult to say. The expression 'new aspect' is vague. It means that we now look at the matter differently—but the question is: what is the essential, the *important* manifestation of this 'looking at it differently'?

First of all I want to say: "It need never have *struck* anyone that in certain products all the factors are equal"—or "'Product of all equal factors' is a new concept"—or: "What is new consists in our classifying calculations differently". In exponentiation the essential thing is evidently that we look at the *number* of factors. But who says we ever attended to the number of factors? It *need* not have struck us that there are products with 2, 3, 4 factors etc. although we have often worked out such products. A new aspect—but once more: what is *important* about it? For what purpose do I use what has struck me?—Well, first of all perhaps I put it down in a notation. Thus I write e.g. 'a^2' instead of '$a \times a$'. By this means I refer to the series of numbers (allude to it), which did not happen before. So I am surely setting up a new connection!—A connection—between what objects? Between the technique of counting factors and the technique of multiplying. . . .

But the *same* proof as shews that $a \times a \times \ldots = b$, surely also shews that $a^n = b$; it is only that we have to make the transition according to the definition of 'a^n'.—

But this transition is exactly what is new. But if it is only a transition to the old proof, how can it be important?

'It is only a different notation.' Where does it stop being—just a different notation?

Isn't it where only the one notation and not the other can be used in such-and-such a way?

The answer to this question is unexceptionable, yet we seem left with the question nonetheless: *for* what can the exponentiation-notation be used that the multiplication cannot? Could we not avoid entirely the use of this notation and say everything with multiplication signs?

We cannot. Once we have use for the concept 'equal factors' we must introduce variables where constants alone dwelled. We want to look at 'n' in '$a^n = b$' not as a metalinguistic sign telling us how many times the symbol 'a' is to be repeated when the equation is written in primitive notation, but as a variable ranging over natural numbers. *This* is the "new aspect" that truly introduces a new concept into mathematics, and not something psychological ("seeing the product in a new light," etc.). In order to define exponentiation so as to take care of the case in which the exponent is a variable, we must make use of the "iterates" as before. Following Quine (*STL*, p. 80), we define $z^{|y}$, i.e., the yth iterate of the relation z. "It is the relation of h to k where, for some sequence w, h is $w'y$ and k is $w'0$ and each succeeding thing "in" the sequence w bears z to the thing before it. That is, succinctly,

$$z^{|y} = \{\langle h, k \rangle : (\exists w)(w \in \text{Seq} . \langle h, y \rangle, \langle k, 0 \rangle \in w . w \mid S \mid \breve{w} \subseteq z)\}.\text{"}$$

This done we can define addition, multiplication, and exponentiation in short order:

$$x + y = S^{|y}(x), \; x \cdot y = (\lambda_z(x + z))^{|y}(0), \; x^y = (\lambda_z(x \cdot z))^{|y}(1).$$

The direct definition of exponentiation, therefore, is as follows:

$$x^y = \lambda_z \lambda_w((S^{|w}(x))^{|z}(0))^{|y}(1).$$

This definition fulfills perfectly Wittgenstein's requirements: to prove, say, that $16^{15} = \ldots$ where exponentiation is as we define it, it is not sufficient to multiply 16 by itself a certain number of times. We must rather construct a sequence whose first member is 16, such that each member is the previous multiplied by 16, and look at the 15th (more strictly, according to our definition, the first member is 1, and we must look at the 16th). Thus the proof that $16^{15} = \ldots$ will indeed "show that I take the number as a factor

Mathematical Knowledge

15 times." And it is indeed true as well that the exponentiation "sets up a new connection between . . . the technique of counting factors and the technique of multiplying." But what Wittgenstein fails to appreciate is that this "new connection" can be expressed in the set theory itself, since the latter quantifies over sequences. This point is the same as the one I made against Parsons. For the latter, too, underestimated the power of set theory to capture important mathematical concepts, and therefore relegated these concepts to the metalanguage. But although all concepts such as the closure of the natural numbers and exponentiation may begin as metalinguistic (e.g., exponents as abbreviations for repeated multiplication signs) they soon develop beyond that stage, as soon as variables are introduced into formulas containing the new symbols.

To prove that $16^{15} = \ldots$, it is not sufficient to perform the calculation $16 \times 16 \times 16 \times 16 \times 16 \times 16 \times 16 \times 16 \times 16 \times 16 \times 16 \times 16 \times 16 \times 16 \times 16$. But equally it may not be necessary. The definition of the exponent does not require it; all that is required is to prove the existence of a certain sequence. One may use various shortcuts one can prove to be efficacious. For example, from the definition of the exponent, one can easily show by induction that

$$(a^m)^n = a^{mn}$$

Hence one can show that 100^2 equals 10^4 by noting that $2 \times 2 = 4$ and not much more; no actual multiplication of factors need be performed, which might be messy in primitive notation.

Returning to our original question, whether a number like 3,455,555,557,575, even written as a polynomial in '10', might not be unsurveyable, because of the necessity to write out in terms of the 'plus' sign alone terms like 3×10^8, we see that the picture is not so grim. In near-primitive notation, 3×10^8 looks like this:

$$\lambda_z(3 + z)^{|\lambda_z(\lambda_w(10 + w)|z(0))|8(1)}(0)$$

where the reader may make the appropriate substitutions for the

Logic and Mathematical Knowledge

decimal numerals. I venture to say that anyone with training and intelligence could do decimal calculation in Russell's system.

We still have not got our calculation truly into primitive notation. We persist in using the notation for iterates. We use symbols to abbreviate open sentences with free variables. We allow ourselves the use of the lambda sign, itself to be removed in favor of the definite description sign, which in turn is explained in context as Russell urged. If we went so far as to eliminate the definite description sign, then, our theorems and proofs would indeed be hopelessly complex.

But this unsurveyability is of different color from the kind Wittgenstein claims exists in proofs proffered in *Principia*. Wittgenstein suggests that many theorems (e.g., those concerning the quantifier-free arithmetic of large numbers) require, in *Principia*, unsurveyable proofs whose validation themselves requires arithmetical knowledge. This, because the proofs will necessarily contain large amounts of repetitive symbolism corresponding to the size of the numerals of decimal calculations of the same facts. "It is not logic—I should like to say— that compels me to accept a proposition of the form $(\exists\quad)(\exists\quad) \supset (\exists\quad)$, when there are a million variables in the first two pairs of brackets and two million in the third. I want to say: logic would not compel me to accept any proposition at all in this case. Something *else* compels me to accept such a proposition as in accord with logic" (II, 16).

Wittgenstein refers here to the first-order version of $1{,}000{,}000 + 1{,}000{,}000 = 2{,}000{,}000$; written out in primitive notation, it would take the millions of variables Wittgenstein has in mind. To obtain first-order equivalents of arithmetic identities, one first defines $(\exists_n x_n)Fx_n$ for each numeral written in place of 'n', as follows:

$$'(\exists_0 x_0)Fx_0' \text{ is } '(x_0) - Fx_0'$$
$$'(\exists_{n+1} x_{n+1})Fx_{n+1}' \text{ is } '(\exists x_{n+1})(Fx_{n+1} \ \& \ (\exists_n x_n)(Fx_n \ \& \ x_n \neq x_{n+1}))'.$$

Then to get the equivalent of $1{,}000{,}000 + 1{,}000{,}000 = 2{,}000{,}000$ one writes:

$$(\exists_{1000000} x_{1000000}) F x_{1000000}$$
$$\& (\exists_{1000000} x_{1000000}) G x_{1000000} \& (x) - (F x \& G x)$$
$$. \supset .$$
$$(\exists_{2000000} x_{2000000})(F x_{2000000} \vee G x_{2000000}).$$

Wittgenstein here and everywhere misrepresents the form of these propositions, since he does not recognize the role of identity in the first-order version of arithmetic.

If one wrote this sentence out in primitive notation, it would be quite out of the question to verify it directly. The only way we could know it to be true is to use the fact that the antecedent contains two groups of 1,000,000 variables each, and the consequence 2,000,000 (more is needed, of course), and our knowledge that $1,000,000 + 1,000,000 = 2,000,000$. This is the kind of unsurveyability that Wittgenstein has in mind. In order to conquer it, there is no choice but to consider the '1,000,000' in '$(\exists_{1000000} x_{1000000}) F x_{1000000}$' a real numeral, and not, as is sometimes claimed, an arbitrary marker that "happens" to be the 1,000,000th member of a series. On the contrary, the essence of the subscripts that garnish the backwards 'E' is not merely to be different from one another (so that we can introduce an infinite number of abbreviations) but to mark the *number* of differing variables which are understood. The difference between the two points of view is that according to the latter we may apply the laws of arithmetic and read off results in logic based upon arithmetic operations performed on the subscripts. Again, once we decide to grant full numeralhood to the subscripts in '$(\exists_n x_n) F x_n$,' such an expression becomes more than an ordinary defined expression—it becomes a metamathematical definite description. We have thus introduced mathematical concepts and knowledge as preconditions for recognizing as true some of the object-language theorems: those we can comprehend only under descriptions that presuppose such concepts and such knowledge.

To put the matter yet another way: the abbreviations of the form '$(\exists_n x_n) F x_n$,' being infinite in number and becoming indefinitely long, must be used in ways that set them apart from ordinary abbreviations, when we try to prove things like the first-order version of '1,000,000 + 1,000,000 = 2,000,000'. Ordinary abbreviations are marked by the following trait: once we substitute definiens for definiendum, we may continue a proof as though the new expression, containing abbreviations, were primitive notation. For example, given the definition of the multiplication sign, '×', we can operate with it as if it were a primitive function sign. Sometimes, of course, the rules by which we manipulate the defined expressions are not primitive rules—take, for example, the contextual introduction of definite descriptions, which allows us to introduce the derived rule, UI—but still, the manipulations themselves are clearly logical. (Often, they may be primitive manipulations in some other system.) Once the substitutions are made, then, we may continue relying on skills of the same kind as we did before the substitutions. To work with the newly defined expressions requires no extra "knowledge that."

The abbreviations '$(\exists_n x_n)Fx_n$' are not in this category. They cannot be pushed around like primitive notation. For what replaces 'n' in '$(\exists_n x_n)Fx_n$' must be regarded as a numeral with the usual role numerals play in calculations. We must still appeal to extralogical truths in proving truths of the form

$$[(\exists_n x_n)Fx_n \ \& \ (\exists_m x_m)Gx_m \ \& \ (x) - (Fx \ \& \ Gx)]$$
$$\supset (\exists_{m+n} x_{m+n})(Fx_{m+n} \lor Gx_{m+n})$$

where n and m are in the billions, *even* after the steps in the proof have been sullied with the formulas '$(\exists_n x_n)Fx_n$'. If the reduction of arithmetic to logic were nothing but the replacement by first-order truths of quantifier-free arithmetic identities, Wittgenstein would have a strong case for denying that this reduction provides a rational reconstruction of knowledge that we actually possess, such as the unshakable certainty that 1,000,000 + 1,000,000 = 2,000,000. And Wittgenstein seems convinced that this view of reduction is correct, and that his criticism is decisive (II,4):

Mathematical Knowledge

Suppose we proved by Russell's method that $(\exists a \ldots g)(\exists a \ldots l) \supset (\exists a \ldots s)$ is a tautology; could we reduce our result to $g + l$'s being s? Now this presupposes that I can take the three bits of the alphabet as representatives of the proof. But does Russell's proof shew this? After all I could obviously also have carried out Russell's proof with groups of signs in the brackets whose sequence made no characteristic impression on me, so that it would not have been possible to represent the group of signs between brackets by its last term.

But Wittgenstein is completely wrong here. What in fact replaces '$g + l = s$' in set theory is not its first-order pale reflection, but the second order '$S^{l}g = s$'. And what replaces numerals like 234,555,667 are polynomials in '10', that is '$SSSSSSSSSS0$', where the operation signs for exponentiation, multiplication, and addition are defined in terms of iterates. One need not go through the laborious proof, then, of

$$[(\exists_{1000}x_{1000})\,Fx_{1000}\ \&\ (\exists_{1000}x_{1000})\,Gx_{1000}\ \&\ (x) - (Fx\ \&\ Gx)] \supset (\exists_{2000}x_{2000})(Fx_{2000}\ \vee\ Gx_{2000})$$

to know that $1000 + 1000 = 2000$. And though it is true that the kind of abbreviation that is necessary (especially the introduction of defined singular terms, one in terms of the other) creates a gross gap between the unsurveyability of the primitive notation equivalent and the finished product, I hope to have convinced you that there is a great difference between the kind of inductive abbreviation that Wittgenstein criticizes and the simple kind of definition that goes on when, for example, we define '$x \cdot y$' as '$(\lambda_z(x + z))^{|y}\,(0)$'. This latter kind does not presuppose, for its effective and practical use in shortening proofs, any truths of mathematics in a blatant way as do the abbreviations '$(\exists_n x_n)Fx_n$'.

Not that there are no problems in understanding the epistemic role of garden-variety definitions. There is a grand controversy between Quine and Wittgenstein whether in definition we enjoy a "double life" or whether the defined expression "has a life of its own." I shall return presently to that controversy. But the controversy does not concern mathematics in particular and its alleged reduction to logic. It is a controversy about logic itself. For any logical theory at all spun out of any logical axioms whatever would have to rest upon definitions if it is to be useful. What gives the attack on the reduction of *mathematics* to logic added plausibility is the false suggestion that the reduction is circular, a notion in turn built on two mistakes that we have now exposed:

(a) the mistake that arithmetic identities are reduced to first-order truths that can be grasped only by introducing an infinitude of defined quantifiers, and

(b) the mistake that the decimal numerals are in set theory replaced by numerals of the form '$SSS \ldots S0$', themselves only manageable by creating the concept of an arbitrary numeral '$S^{(n)}0$'; together with the related mistake that exponents in their set-theoretical interpretation are regarded merely as abbreviations for repeated occurrences of '$\times$', the multiplication sign. In general, then, the mistake is thinking that to construct mathematics a special kind of abbreviation is necessary, a two-faced denizen of the object and metalanguages—abbreviations like '$S^{(n)}$', for n arbitrarily large; '$(\exists_n x_n)Fx_n$'; and 'x^n', where this is erroneously regarded as an abbreviation for $n - 1$ occurrences of the multiplication sign.

As we have shown, however, arithmetic is carried out in set theory, not first-order logic; the largest numeral '$S^{(n)}0$' necessary to do decimal calculation is '$SSSSSSSSSS0$'; exponentiation, addition, and multiplication signs are all defined in the system in terms of iterates.

Even after their refutation, however, Wittgenstein's arguments retain force. They fail to refute logicism, but they might refute a view often associated with logicism: that identities, of the form '$a + b = c$' and perhaps others, can be "reduced" to first-order logic with the help of the numerical quantifiers (see, e.g., Putnam, 1967). This view may be wrong; the identities would be reduced to unsurveyable, unlearnable, formulas. But one would have to delve more deeply than I have in these pages on the relationship between mathematical knowledge and language to decide the matter conclusively. My concern has been merely to point out that *even if* Wittgenstein is right about first-order logicism, he loses the argument with Russell. As long as second-order logic, or set theory, provides surveyable reductions of arithmetic truths, logicism is safe from Wittgenstein.

III

We have disposed of the idea that the arithmetical propositions themselves are needed to acquire knowledge of their set-theoretical stand-ins, by showing that these stand-ins are not, as Wittgenstein claims, the first-order truths which involve the numbered quantifiers; but rather second-order truths much more easily taken in. Again, we have scotched the notion that multiplication is defined inductively as repeated addition, that the multiplication sign is a short form of writing many '+' signs.

It is still true, however, that, written in primitive notation, many elementary propositions of arithmetic come out too long to be taken in by the eye. This is true even if we assume that the lambda notation is primitive. Definitions are required to make even the simplest theorems and proofs readable—even if not inductive definitions, the kind that comprehend an infinity of cases. Function signs must be introduced; new predicates imposed upon the language ("is a sequence," "is a function"). The most we seem able to do is to show that these defined symbols are eliminable and that for each "shortened" proof there is a primitive proof. But then some of Wittgenstein's complaint retains force: "One need not acknowledge the Russellian technique at all—and can prove by means of a different technique of calculation that there must be a Russellian proof of this proposition. But in that case, the proposition is no longer based upon the Russellian proof." Or again: "By making it possible to command a clear view of the Russellian proof, I prove something about this proof" (II, 14). Wittgenstein does not say, but the only "technique" of calculation that does the trick is mathematical induction on the length of proofs. Thus, Wittgenstein's objection (once purged of the objections exposed in the second section of this chapter) collapses into a Poincaréan objection (suggested to me by Baruch Brody). Let us be clear on what the objection is, and what it is not.

First, Wittgenstein is not claiming that, to *determine* that logic could form an adequate basis for mathematics, we rely on mathematics. For although this is true, it is irrelevant to logicism. If this were the objection, it would be open to the same rejoinder we made to Poincaré (see above, pp. 29–30). Logicism makes no claim about its own verification, nor does logicism have to be known true in order to *be* true. To use mathematical induction in showing that all proofs of mathematics can be reproduced in logic is not a disproof, but if anything a proof of logicism.

Wittgenstein's objection is rather this. Logicism is, among other things, an epistemological theory. It holds that mathematics is— epistemologically—based upon logic. This means partly that it is possible to know all of mathematical knowledge by knowing only logical truths and logical rules. It also means that logical truths alone show how mathematics can be known a priori. It is a necessary condition, that is, of knowing a mathematical truth a priori that one know a proof of the truth in some logical system. (This assumes that mathematical knowledge requires proof, a view that in fact I reject, though one that classical logicists have held. One might say that the condition should read: In order to know a mathematical truth a priori, one must know its logical representation whether by proof or otherwise.) Few logicists would insist today on a particular logical system, but the existence of at least one logistic system in which mathematical truths could be represented is a *sine qua non* of their position.

Since, then, *we* can certainly know mathematical truths, it must be possible for us, if the logicist is right, to know the logical truths of which arithmetic truths are allegedly abbreviations, and hence to come to know them (see the Introduction). But the logical truths are not all self-evident; they must be proved. And if the proofs are unsurveyable, as most will be, we *cannot* come to know them by their proofs. The only recourse, therefore, is at least to *show that* proofs *exist*; but that would require mathematics, and we have been assuming no independent knowledge of mathematics. Conclusion: either we do not know very much mathematics or else logicism is false.

Wittgenstein's objection differs from Poincaréan objections to logicism in two ways.

(a) Poincaré saw mathematical induction as the ground of our knowledge of mathematical *generality*. Wittgenstein, on the other hand, is speaking equally of singular truths, such as

$$1,000,000 + 2,000,000 = 3,000,000$$

Even such truths, viewed as mere abbreviations of logical truths, cannot by us be learned or justified through direct construction of particular proofs, although they exist. Even here (Wittgenstein would say) one must stoop to mathematical induction to show that a proof exists in logic, from the hypothesis that it exists in arithmetic. (The logician may protest that for singular propositions induction is not necessary, but any other kind of proof would undoubtedly involve us again in unsurveyable calculi.)

(b) Similarly, Poincaré's point does not rely on the limitations of our reason, as Wittgenstein's *seems* to (I think that Wittgenstein would not like this characterization of his argument). Where mathematical induction is for Poincaré necessary and basic, it is necessary for any intelligent being, of arbitrarily large spanning powers. Poincaré's philosophy is not based on the "contingency" that *we* know mathematics, but amounts to an a priori rejection of the logicist position, based on an insight into what mathematical knowledge is, or perhaps into the concept of natural number.

Wittgenstein's argument is this. One can, he admits, prove

theorems in an "abbreviated calculus," a calculus into which the everyday symbols '×', '+', etc., have been introduced by definition. But what is involved in seeing that, in such cases, there *must* be proofs, expressed only in primitive notation, of the same theorems? Wittgenstein thinks that mathematical knowledge is obviously required, and we agree that *if* it is, *then* much of the logicist position is refuted, with the value of what remains open to question.

In assessing Wittgenstein's ideas, one must never forget the distinction already emphasized here between knowing *how* and knowing *that*—after all, we are studying the philosopher who popularized the distinction. Knowing how to "follow a rule" is knowledge only in the sense of having certain abilities—even if the rule itself is inductively defined. In discussing what is needed to know the existence of primitive proofs on the basis of "abbreviated" proofs, one is interested only in "knowing that," or "propositional" knowledge. Logicism need not commit itself to any position concerning the *abilities* and *skills* presupposed in mathematical knowledge. The question remaining, therefore, is this: Is arithmetic knowledge (knowledge of arithmetic truths) presupposed in coming to know a theorem on the basis of an "abbreviated" proof?

Recall the abbreviations we are discussing. They are not those that introduce symbols like '$S^{(n)}$' or '$(\exists_n x_n)$'—inductive definitions. These (were they crucial) would, indeed, present problems, as we saw in the previous section. And in that section, therefore, we were at pains to show that this kind of abbreviation need not be used at all. The kind of definition needed, rather, is exemplified above, where we defined '+' or '×' using the lambda notation. Abbreviations are needed that replace singular terms with free variables by defined singular terms containing the same variables. Definitions of this type do not change the structure of the sentences that make up the proof. Each defined symbol is just a reminder of longer text.

Mathematical Knowledge

Now Wittgenstein's objection bases itself on the premise that a proof into which defined symbols have been introduced cannot itself be called a logical proof; the real logical proof can be only the "expanded" version. To the extent that *logic* convinces us of the truth of some statement, the convincing may be accomplished only through the primitive notation and none other. This view is conveyed by characterizing symbols like '$\times$' or '$+$' in a formal system of logic as "abbreviations," a characterization to which we have so far acquiesced.

Logicists, admittedly, have often viewed the introduction of defined symbols into formal systems in this way. Thus Church: "Abbreviations . . . are not part of the logistic system . . . but are mere devices for the presentation of it" (p. 75). Church goes on to ask the reader, whenever the latter sees an abbreviation of a well-formed formula, "to pretend that the well formed formula has been written in full and to understand us accordingly." And he challenges the reader who balks at this "to rewrite this entire book without use of abbreviations, a lengthy but purely mechanical task" (p. 76). And Quine, for many purposes also a logicist, agrees that definitions in a formal system "are best looked upon as correlating two systems, two notations, one of which is prized for its economical lexicon and the other for its brevity or familiarity of expression" (*WP*, p. 112). Of course it is just this view that leaves the logicist open; for in most cases it is just impossible to "pretend," along with Church, that the primitive version of an "abbreviated" expression has been "written in full."

But there is another kind of definition (Church, p. 76; Hilbert and Bernays, vol. 1, pp. 292–293 and 391–392). To quote Church again (p. 76n.), definitions of this kind "are intended to extend the [object] language by introducing a new notation not formerly present in it." Syntactically, definitions of this type look like what Church there calls "explicative definitions"; that is, they have the appearance of "object-language" statements—they use, rather than mention, linguistic objects, and are usually either identities

or material biconditionals. But explicative definitions "are intended to explain the meaning of a notation . . . already present
in a given language." Stipulative definitions (as we shall call
them), however, are not, formally, a part of the original language,
for they contain a "new notation" that they introduce. Stipulative
definitions are thus really part of the expanded object language.

"Introduction," here, is something of a metaphor: if the rules
for stipulative definitions are set out fully, they fix in advance
every such definition possible. Each language that permits stipulations, therefore, is really an infinite set of object languages with
a common core, differing only as to how references or extensions
are to be assigned via stipulation to the "extra" terms. "Introduction," however, has an epistemological connotation: a stipulative definition—if an identity or biconditional—could be used to
teach the use of the defined term in logic and mathematics. A
person who uses a term in accordance with a stipulative definition
has learned or resolved never to use the term in any way that does
not stem directly from the convention. Stipulative definitions are
therefore unlike so-called "analytic" truths, which are, if they
exist at all, discovered rather than legislated, for the latter truths
allegedly depend on established usage, while the former establish
usage.

Stipulative definitions are indeed "true by convention," and so
knowable a priori. Nevertheless, they need not be "necessary"
truths, again unlike "analytic" truths. (I rely here on lectures of
Saul Kripke; for the sake of argument, I assume the concept of an
analytic truth to be clear, since the sequel does not depend on that
assumption.) This is clearly so if the object language contains
definite descriptions and proper names to begin with. For example: suppose we wish to name the color of Jack's car—whatever
color it may be. If we stipulate

x is *Kolor* if and only if x has the same color that Jack's car
in fact has. (Df.)

we have *truth* by stipulation. But we do not have necessary truth; the statement is not analytic. This is not merely because Jack's car's existence is not necessary; for even the truth

If Jack's car exists, x is *Kolor* if x has the same color as Jack's car

is not necessary. Consider the "possible world," in which Jack's car exists but is colored magenta—a different color, surely, from the color it has in fact. Since '*Kolor*' is defined to denote the color it has in fact, we have in the aforementioned possible world, for some x

x is *Kolor*—true
x has the same color as Jack's car—false.

Though we know a priori that if Jack's car exists, then it is '*Kolor*', it is only a contingent matter that it has that or any other color. Again we see that truth by stipulation is not the same as analytic truth.

Stipulative definitions fix the reference of a singular term, the extension of a general term, without necessarily fixing "meaning." Some may argue that the kind of counterexample I have adumbrated above fails in the case of mathematical entities; for the success of the counterexample flows from the contingency of Jack's car's color, but mathematical entities have only essential properties (thus, e.g., Kripke). If so, stipulated truths of mathematics would be, not only a priori, but necessary. Not a worrisome prospect, concerned as I am right now with *defending* the logicist program from the attacks of such as Wittgenstein. Even, however, should there be contingent mathematical properties, the logicist need not worry. For he is after the secret of mathematical knowledge, the grounds for accepting mathematical truth. And stipulative definitions, contingent truths though they may be, are nevertheless true by convention, and hence knowable a priori.

But has not Quine disposed of truth by convention? Quine (1936) indeed shows that conventionalism is false for elementary

Logic and Mathematical Knowledge

logic (truth functions and quantifiers). The proof depends upon the infinity of conventions necessary to make elementary logic true by convention; but the logicist, as I represent him, does not need so many conventions as that. (True, the logicist will have a harder time explaining why logic is an a priori science, and hence why mathematics is ultimately a priori, in the face of Quine's critique. But in this section, I am dealing only with the claim of logicism that mathematics is "based on" logic, and not with the further problem of whether set theory is logic and known a priori. I am attempting to show that whatever set theory and logic are, the logicist may claim that we may get from them to mathematical knowledge via a priori truths, stipulative definitions, and so that the problem of surveyability, at least, is unreal.) And despite his strictures against *analytic* truth, Quine agrees that there is nothing problematic about stipulating a certain statement as true, where the statement is used to define a new term. (N.B. we shall deal immediately with Quine's thesis of the "passing efficacy" of stipulation.) In fact, he seems to hold that even the axioms of set theory, or some of them, are conventional, when he says: "Comparative set theory has now long been the trend; for, so far as is known, no consistent set theory is both adequate to the purposes envisaged for set theory and capable of substantiation by steps of obvious reasoning from obviously true principles. What we do is develop one or another set theory by obvious reasoning, or elementary logic, from unobvious first principles which are set down, whether for good or for the time being, by something very like convention" (*WP*, p. 104). Or later, in the same essay: "In set theory . . . convention in quite the ordinary sense seems to be pretty much what goes on. Conventionalism has a serious claim to attention in the philosophy of mathematics, if only because of set theory" (p. 108). Note, however, that our doctrine is quite a bit more modest than the view that the axioms of set theory are true by convention. For we are speaking only of the identities and material biconditionals of unmistakable *definitions*, which introduce novel notation, such as our familiar:

$$x + y = S^{|y}(x) \, (\text{Df.}).$$

Quine would surely go along here at least with the view that such sentences *might* be *introduced* into a language as true by stipulation. This is the only point I wish to make initially.

But is it really true that the logicist's definitions are *stipulations*? Can we really say that '$+$' is an arbitrary symbol with no pre-established use? Is it not rather, that a definition like

$$x + y = S^{|y}(x)$$

is no arbitrary stipulation, but an exercise in *explication*? (In addition to Church, p. 76n., cf. Quine, *FLP*, p. 25) And this type of definition clearly relies, if anything, on "prior relations of synonymy" (Quine, *FLP*, p. 27). Quine should object, then, to definitions like the above one, just as he objects to synonymy—*if* such definitions are intended to express a priori *truth*. (Of course, he does not object to explication as such, only to the claim that explication is the finding of "meaning," the discovery of truth by virtue of such preexistent meaning.)

Luckily for the logicist, this objection is based on a now-familiar confusion: the confusion between what logicism says and how logicism is known. As I see it, logicism does not uncover the old grounds of knowledge in mathematics; it does not assert that our grounds were always logic. Rather logicism is the view that mathematics *can* be grounded in logic. Indeed, the logicist, in demonstrating the correctness of his philosophy, must explicate familiar mathematical concepts in terms of logic. He must create a logical system that can supersede the axioms of mathematics. He must capture the ordinary usage of the numeric discourse, so far as ordinary usage is worth preserving—and, by and large, it is. His success depends upon his ability to explicate mathematical terms in the idiom of logic. The philosopher, the logicist, must make do with the preexisting use of mathematical terminology, the "preexisting synonymies."

But none of this enters into what logicism is about, what logicism says. Logicism is a claim about what a *mathematician* can do. Logicism claims that a mathematician can learn mathematics from scratch, knowing only logic to begin with. From that mathematician's point of view, definitions such as

$$x + y = S^{|y}(x) \,(\text{Df.})$$

may be pure stipulation, even if the framer of the system chose the definition for stipulation on the basis of established usages. The claim is not that *logicism* is true by stipulation, but that certain definitions may be considered as stipulations by a mathematician who wishes to know mathematics purely a priori. Such a mathematician, the claim goes, need not know the established use of '+'. He need learn only

$$x + y = S^{|y}(x)$$

as a legislated truth, completely unaware that this definition has been chosen by someone anxious to explicate '+' as it appears in ordinary arithmetic. In general, the mathematician will lose nothing (so claims logicism) by learning *nothing* but

(1) the axioms and rules of logic
(2) various definitions, stipulated as true.

And since knowing truths (1) and (2) is knowing "mathematics" a priori (continues logicism), maybe we all ought to stake our mathematical knowledge on logic (this second claim will be taken up shortly). Why these, rather than other definitions, were chosen is not important to the mathematician; it is only important to the logicist, in proving that with *these* stipulations the mathematician can perform the whole repertoire of mathematics, leaving out nothing of importance in that repertoire.

In sum: definitions add to a vocabulary, but need not lead their user out of the a priori. We may grant our mathematician this very weak ability to learn new terms, terms that we can prove eliminable, though the mathematician cannot eliminate them on pain of unsurveyability. Stipulative definitions behave like axioms. Proofs containing such "axioms" will now be surveyable, and if logical axioms are knowable a priori, so will all of mathematics as generated by the logistic system. The "abbreviated" proof becomes the real proof, the primary vehicle of knowledge, for it is in reality not abbreviated at all—definitions are not abbreviations.

We depart thus from Russell's view, that in a definition such as

$$x + y = S^{|y}(x) \, (\text{Df.})$$

"the sign of equality and the letters 'Df.' are to be regarded as one symbol, meaning jointly 'is defined to mean'" (Van Heijenoort p. 169). According to this, definitions are again statements about language, albeit disguised in a way that encourages use-mention confusion. In later logical work, we find symbols such as '$=_{df}$' to make Russell's point even clearer. My view is that '$=$' is univocal: it always expresses identity, the relationship which each entity bears to itself and no other; but 'Df.' marks a special way of asserting a statement, marks a stipulation. (This seems also to be Frege's standpoint.) Writing 'Df.' aside a statement serves a function similar to that of writing 'Axiom' aside an asserted proposition: both excuse the mathematician from further justification. Logically, of course, both 'Axiom' and 'Df.' are superfluous, since ideally whether a formula is an axiom or a definition can be effectively checked, just as we can effectively check whether a line of a putative proof follows from preceding lines by a rule of inference. Anyhow, 'Df.' had better be superfluous: our envisioned mathematician does not have that word in his vocabulary at all, though he certainly can recognize a definition when he sees one, in the sense of knowing what to do with it.

Logic and Mathematical Knowledge

As promised, we must now deal with Quine's warning about conventional truth:

> Conventionality is a passing trait, significant at the moving front of science but useless in classifying the sentences behind the lines. It is a trait of events and not of sentences.
>
> Might we not still project a derivative trait upon the sentences themselves, thus speaking of a sentence as forever true by convention if its first adoption as true was a convention? No; this, if done seriously, involves us in the most unrewarding historical conjecture. Legislative postulation contributes truths which become integral to the corpus of truths; the artificiality of their origin does not linger as a localized quality, but suffuses the corpus. If a subsequent expositor singles out these once legislatively postulated truths again as postulates, this signifies nothing; he is engaged only in discursive [i.e., explicative] postulation. He could as well choose his postulates from elsewhere in the corpus, and will if he thinks this serves his expository ends.   [*WP*, pp. 112–113]

But I do not think the warning relevant to the stipulations I have in mind—the definitions that introduce mathematical symbols into the object language. Definitions, though they look like them, are not postulates of logic. The postulates of set theory may change, thus changing the significance of the epsilon and thus of all the defined terms. But the change is global and uniform, leaving

$$x + y = S^{|y}(x)$$

true by convention as before. Furthermore, it is reasonable, indeed almost certain, to assume that '$S^{|y}(x)$' will be definable in the face of postulate change. The reason is, of course, that the postulates of logic are selected partly for their ability to support such concepts. ($S^{|y}(x)$ is defined using the ancestral, an indispensable concept in set theory, whatever the postulates.)

We can thus assume that defining '$x + y$' as '$S^{|y}(x)$' will remain *possible*. Can we be sure, however, that this definition will never become *outmoded* with the progress of mathematics? Can we be sure we will not need a new definition of addition in order to prove arithmetic theorems as yet undiscovered? (After all, there are an infinite number of truths involving addition that are *not* provable from the Peano postulates, but that *are* provable in ZF set theory, to say nothing of future extensions of set theory. For a discussion of the significance of the undecidability of some arithmetic propositions even in set theory, see Benacerraf, 1960.) Yes we can. (I owe many of the following ideas to a conversation with Leslie Tharp.) For whatever we learn about addition, we will never *change* these two basic truths:

$$x + 0 = x$$
$$x + Sy = S(x + y)$$

both of which are provable via set theory from our definition of addition, assuming some standard definition of 'S'. But in basic set theory (the part that may be added to, but will never, it seems, be diminished) it is easily shown that only one function on the domain of natural numbers satisfies these two equations. No change, then, in the definition of '$+$' can make any difference in the numeric theorems provable in set theory as it is now, or in any extension. This, as long as S itself remains fixed. But it is obvious that no gain can come of redefining 'S', even in the face of an extension of set theory to encompass more arithmetic truths. Though this hardly needs argument, an argument is available: it is again easily shown in set theory that the natural numbers, as usually defined, (the S-descendants of '0', where 'S' denotes any one of the commonly used successor relations) with the relation S, constitute a model for *second*-order Peano arithmetic (Church's A_2, p. 321), as well as first-order arithmetic (Mendelson's S, p. 103). Furthermore, any acceptable set-theory version of arithmetic seems bound to model second-order arithmetic too. The second-order postulates do seem to capture the central concepts of arithmetic. First, second-order arithmetic is categorical, and thus semantically *complete* ("complete as to consequences," Church, p. 329 and **555, p. 330). Second, the first-order postulates are infinite in number, owing to the axiom schema of induction. And it seems that the evidence we have for the schema derives rather from our grasp of the unitary second-order induction axiom, than from our fortuitous individual grasp of the instances of the schema. Finally, the second-order axiom "says more" than the axiom schema—for it mandates induction upon arbitrary properties of numbers, and not merely the denumerable properties that happen to be expressible in the vocabulary of Mendelson's S. So it is entirely reasonable that any model of arithmetic, present or future, in set theory should model *second*-order arithmetic.

But the second-order Peano postulates are not only categorical; they are provably categorical. Let N and S be the protagonists of our present reduction of number to set theory, with N' and S' the protagonists of any other; let A be some proposition of set theory relativized to N and containing 'S'—and A', the same proposition relativized to N' and with 'S' replaced by 'S''. We can prove, with the help of elementary set theory (i.e., containing no questionable propositions that might conceivably be overturned by progress, such as the axiom of choice, the axiom schema of replacement, etc.) that A *if and only if A'*. That is, A is provable in set theory (in any conceivable set theory that might be ever accepted) if and only if A' is provable in that set theory. This is overwhelming evidence that we need never change any of the conventions according to which logicism introduces arithmetical concepts by definition into set theory, in the face of arithmetical progress. These conventions are thus *exceptions* to Quine's doctrine.

Quine's position, after all, is really this: not only is it "un-historical" to imagine that a stipulated truth remains stipulated for any length of time, but it would be *irrational* for a scientist to *hold* a scientific statement forever true by convention. (Though he could, if he were so foolish.) Quine's appeal to "history" supports his normative claim. The logicist's position, as I see it here, is that Quine is wrong about certain definitions in mathematical logic. Once discovered, there is no reason to think they might *have* to be changed some day, on pain of being stubbornly unscientific.

Every definition has its rivals. Every convention, including arithmetical conventions, *could* be changed. The logicist simply reconciles himself to the multiplicity of possible reductions. *Each* reduction is a way of getting along in mathematical contexts by knowing only logic—getting along better than usual, because logic is an a priori science. (Or so the logicist says. The next section will examine this assumption. So far we have claimed only that our conventions are known a priori.) Given any single reduction, the logicist position is that it need not be changed, that a mathematician who sticks with a particular reduction is at no disadvantage, that one might hold

$$x + y = S^{|y}(x)$$

as true by stipulation in the future as well as now. The logicist's optimism is based on the Peano postulates, the eternal touchstone for a good reduction of arithmetic to logic, and on elementary model theory (the theorems mentioned above), portions of set theory that are unshakable.

To conclude: The axioms of logic, plus a set of stipulations, are supposed capable of producing all of arithmetic. It is thus possible to come to know arithmetic, as well as the rest of mathematics, by knowing only the axioms and the supplementary definitions. Definitions are knowable a priori, since they are true by convention. *If* the axioms of logic are also known a priori, we see how mathematics can be an a priori science. The definitions, though they expand the object language, add nothing to the expressive power of the object language (see Shoenfield, 1967, section 4.6, for a precise statement of this claim and a proof of it) "in principle," though it would be impossible for a mathematician to attempt to work in an object language without definitions.

But since definitions are part of the object language, part of logic, not an "abbreviation" of logic, Wittgenstein's view that surveyability is impossible in logistic proof is totally false. It is worth repeating that Wittgenstein blundered into this position by unfairly classifying the logicist's definition as inductive, as containing an implicitly numerical variable. This type of definition does produce either circularity or unsurveyability. *These* definitions can by no means be stipulations. For there would be an infinity of such stipulations, impossible for a language of finite beings. (Any attempt to get around the problem must regard the pseudovariable in, say, '$(\exists_n x)$' as a true numerical variable, producing circularity. See the previous section for details). Once we have shown that such definitions play no role in the logicist program, we are left with a residue of definitions that can indeed be treated as stipulations *within* a mathematical language.

The epistemological objections to logicism thus fall away. There is no circularity in the reduction of arithmetic to set theory; if set theory is solidly based, a mathematician can in principle use it, to the exclusion of any other mathematical theory, in coming to know the rest of mathematics, including arithmetic. Any such mathematician will indeed require certain mathematical *skills*, even arithmetic skills, before he can learn arithmetic from set theory. But skills are not knowledge in the cognitive sense at issue here, and many of the epistemological critiques of logicism seem to confuse "knowing how" with "knowing that."

Logicism must therefore be taken seriously as an epistemological theory, despite many of the criticisms of Poincaré and Wittgenstein. Logicism forces us to admit at least that the epistemic status of arithmetic is no worse than that of set theory.

Mathematical Knowledge

To argue further, however, that this shows, for example, that the propositions of arithmetic are necessary requires one to demonstrate this of the propositions of set theory. Similarly, to argue that logicism explains the indubitability of mathematical knowledge is to presuppose the indubitability of set theory.

Only conditional support for logicism, then, can emerge from this chapter. Granted the feasibility of the logicist reduction, the question of its desirability remains. And this is the crucial question to which we now turn.

Two

Logicism Reconsidered

Poincaré and Wittgenstein are not the only critics of logicism, though they are among those few who ever claimed the logicist reduction impossible. Other philosophers grant the reduction, but question its value. Though arithmetic may be reducible to set theory, the latter hardly qualifies as "logic," certainly not if the principles of logic are supposed to be "analytic" (Benacerraf, 1960; Parsons, 1965, pp. 195–197).

These criticisms, devastating to the aims of logicism (unlike the rejected arguments of Chapter One), leave one of its theses intact. This is the view that arithmetic (epistemologically) reduces to set theory, that one could learn arithmetic from set theory, that arithmetic and set theory are parts of the same subject.

Even this view has its critics (Benacerraf, 1960). One may question whether set theory is an adequate foundation for arithmetic, for set theory is often regarded as ailing. Parsons (1965) succinctly lists the symptoms: "the multiplicity of possible existence assumptions, questions about impredicative definitions, the paradoxes, [and] the possible indeterminacy of certain statements in set theory such as the continuum hypothesis" (p. 197;

71

Mathematical Knowledge

Parsons does not draw all my conclusions from this data). Why would anyone *want* to learn arithmetic by doing set theory? Why live in fear of finding oneself in Frege's shoes, of finding the foundation of one's arithmetic "tottering" (Frege's expression, in the face of Russell's paradox) and, with it, arithmetic itself? Does arithmetic need a foundation, and could set theory provide it?

There is still a view akin to logicism which even this argument does not touch (though Poincaré's and Wittgenstein's, if valid, would have)—Quine's. He does not claim that arithmetic truths are analytic (Quine thinks no truths so describable) nor does he argue epistemologically for the reduction. But he does hold the reduction philosophically significant. This significance lies in our ability, through the reduction, to "get by without numbers." The reduction shows that sets can "do the work" of numbers, or even that numbers "are" sets (in a certain sense). To the extent that Quine holds that numbers are sets, as is every other mathematical object, the claim is not, of course, that each particular number is identical with some particular set. It is rather like the claim that the x, the y, and the z axes are lines, though not any particular lines. Benacerraf's objections (1965) do not damage Quine's position—not that Benacerraf ever said that they do.

When Quine suggests that we "get by without numbers," he is promoting the reduction of number theory to set theory. For him, reduction is merely reinterpretation, in this context at least: "A usual occasion for ontological talk is reduction, where it is shown how the universe of some theory can by a *reinterpretation* be dispensed with in favor of some other universe, perhaps a proper part of the first" (*OR*, p. 55, italics mine). "Getting by without numbers," then, is reinterpreting as referring to particular classes, the terms purporting to denote numbers; reconstruing as true of classes, predicates taken to be true of certain numbers; redefining as ranging over classes, variables understood to range over numbers. Since we "need" sets anyhow for other mathematical purposes, the possibility of such a reinterpretation, Quine

Logicism Reconsidered

holds, is grounds enough to support an "ontic" conclusion: that numbers are to be "eliminated," "dropped" from the universe.

Since the reduction of arithmetic to set theory merely models the former in the latter, no change in the arithmetic discourse is made; it survives, totally intact, in its new guise. Indeed, were it not for the assumption of a "background" theory (set theory, in this case) it would make no sense to talk about a reinterpretation, for the arithmetic discourse, in its internal relationships, changes not at all (see Quine, *OR*, pp. 53–55). Quine thus exaggerates when he says (referring to the reduction of the ordered pair to set theory):

> A similar view can be taken of every case of explication: *explication is elimination*. We have, to begin with, an expression or form of expression that is somehow troublesome. It behaves partly like a term but not enough so, or it is vague in ways that bother us, or it puts kinks in a theory or encourages one or another confusion. But also it serves certain purposes that are not to be abandoned. Then we find a way of accomplishing the same purposes through other channels, using other and less troublesome forms of expression. The old perplexities are resolved.   [*WO*, p. 260]

Yet here is a reduction in which what is reduced is unexceptionable. The reduction of arithmetic to set theory does not even provide a relative consistency proof of any value, since we have seen which theory has notorious consistency problems. Nor does the reduction as such add even a single arithmetic truth we could not have obtained without it. For although some arithmetic truths may be proved only with the aid of set theory, that aid is forthcoming without the reduction. Instead, the reduction makes newly meaningful a host of questions such as "Is 3 a member of 259?"—questions now clamoring for answers, dead weight on a starkly simple theory of numbers.

Quine can, however, point to theoretical economies as a motive for reinterpreting number theory and thus "getting rid of numbers." Getting rid of them is a goal in itself: instead of being saddled with two theories, we can make do with set theory.

Mathematical Knowledge

Instead of having to postulate sets *and* numbers, we might as well postulate just the sets. "Frege and von Neumann showed how to skip the natural numbers and get by with what we may ... call *Frege classes* and *von Neumann classes*" (*WP*, p. 200). The ability to "get along" without postulating numbers is by itself a good reason to do so.

Quine is aware, of course, that set theory is threatened by paradox; toward the end of *Word and Object* (pp. 269–270), he even suggests that number theory be kept *along with* a watered-down version of set theory, to avoid trouble. In *Ontological Relativity* (p. 43) he notes:

> It will perhaps be felt that any set-theoretic explication of natural number is at best a case of *obscurum per obscurius* .... I must agree that a construction of sets and set theory from natural numbers and arithmetic would be far more desirable than the familiar opposite. On the other hand our impression of the clarity even of the notion of natural number itself has suffered somewhat from Gödel's proof of the impossibility of a complete proof procedure for elementary number theory, or, for that matter, from Skolem's and Henkin's observations that all laws of natural numbers admit nonstandard models.

Since we have, alas, only a construction of number theory out of set theory, and since it is wasteful to "keep" sets and numbers both, we must (even if with qualms) reduce the numbers to sets. Ontic economy is thus sufficient grounds to "skip the natural numbers." This position will be criticized shortly.

Meanwhile I am unable to follow Quine in his *tu quoque* against the natural numbers. Certainly the *relative* clarity of number theory vis à vis set theory needs no reassessing because of the discoveries Quine mentions, since Gödel's theorem applies just as much to set theory as to number theory. Further, the incompleteness of set theory is notoriously worse: we have no idea at present whether the Continuum Hypothesis is true or false, or even how to decide the issue, though we know that the

Gödel sentence for number theory is true, even if independent, or rather, because it is independent.

To pick up the thread: Quine does grant that set theory is epistemically on shakier ground than number theory. Quine nevertheless would argue that we still need set theory, shakier or not, to conduct most of our business. Why not, then, "skip the numbers"?

This position presupposes that we can divorce ontological from epistemological questions, that one can achieve ontological without epistemological gain, indeed at epistemological loss. This seems to me an absurdity, especially for a pragmatist like Quine. Quine's position is that if one theory, with its implied ontology, can do everything, and more, than another, then we can junk the latter. As to the question of what theories "do" (i.e., *what* can theories do better than others), Quine, as a good pragmatist, answers: They enable us to predict surface irritations of our sensory organs (*WO*, chap. 1). Now *set* theory, *if* it were true, could do as much and more than number theory; but since we are less sure that set theory is true than we are of number theory (owing to potential inconsistencies) we cannot retire number theory—yet. Perhaps Gödel's second incompleteness theorem can be interpreted to say that this situation is permanent; at any rate it exists now.

We are in the situation of the data-processing director of a large firm who is offered a proposal by IBM to exchange his small CDC computer for a more elaborate one. On paper the trade looks good: the flowchart of the IBM includes that of the CDC, so that in theory the former could do the work of the latter and more. The catch: the IBM machine is prone to break down, to slow down, to err. Even if we assume that the director must rent the IBM because it *is* flowcharted to solve problems beyond the range of the CDC, *must* he then throw out the latter? In the business world, indeed, there *are* considerations of cost and of space which might dictate ridding oneself of the CDC, but it is not so clear that similar conditions govern the acceptance of theories.

True, the mathematician might plunge ahead with the reduction of number theory to set theory, reserving the option to "resurrect" the former, should the latter prove embarrassing. But I see no reason to justify such an action. The exigencies of life often do not allow gambling with truth. To revert a final time to the computer example, the CDC may be needed to write payroll checks. It might be dangerous to replace it with the newfangled contraption: people might be left without money at the end of the month, while the data-processing manager is busy "resurrecting" the old machine.

At this point protest is likely that the analogy is not apt. For Quine does not ask that one junk the numerical *calculus* (reply suggested by Benacerraf). In "Definition and the Double Life" (*WO*, pp. 186–190) he says specifically that for all practical purposes one keeps a reduced discourse, while the reduction serves "theoretical" ends such as ontology pruning, simplification, and so forth. The reduction invites one merely to reinterpret the numerical calculus, not to change it in any way. One continues to get the same results as before; they are simply construed in a different universe of discourse. Hence the problem of consistency does not arise for that portion of set theory that consists of the (reinterpreted) Peano axioms and their closure under the rules of inference. (We shall take "closure" here to mean "first-order closure.") This portion of set theory, in fact, is just as consistent as is number theory.

But how do we know this? Or, more to the point, *would* we know it if we rid ourselves hypothetically of the burden of numerical knowledge, which is our grasp of the "standard model" of Peano's axioms? Is there not, then, a clear sense in which set theory cannot take over the job of number theory, from a pragmatic point of view? Will not number theory still be needed to supervise set theory's activities?

Logicism Reconsidered

But, you say, the postulates of set theory needed to do the reduction of number theory (e.g., ZF without the Axiom of Choice and the Axiom of Infinity, but with, say, Quine's 13.1 in STL—which guarantees, among other things, the existence of the first x natural numbers, for every x) are just as certain as number theory; even if we knew no number theory, we could still be sure that this truncated set theory would cause no problems. This argument is hard to beat, since indeed this part of set theory is *modeled*, in fact, by number theory. (Let ϕ be zero, and the set $\{a_1, a_2, \ldots, a_n\}$ be the number $2^{a_1*} \times 3^{a_2*} \times \cdots \times p_n^{a_n*}$, where the a_i* are the correlates of the a_i. The epsilon is then given the obvious meaning.) Still, we *would* lose epistemically if we did not know number theory, even though "absolutely" and "objectively" we would have nothing to worry over. This is no paradox; a person must act according to the evidence *available* to him. No paradox, but is it true?

I think it is: the impredicative nature of set theory, even without the Axiom of Choice and the Axiom of Infinity, makes it less certain than number theory, assuming, as we are, no access to number theory and its reassuring modeling powers.

To illustrate the point, consider Quine's finitistic construction of number theory (STL, sec. 11). The classical infinitistic definition of "x is a natural number" is the Fregean "x is in every hereditary set containing 0"—where a hereditary set contains the successors of all its members. The law of mathematical induction follows directly. We assume

$$F0$$

and

$$Fy \supset FSy, \text{ for all } y.$$

Given further that

$$x \text{ is a natural number,}$$

we must prove that Fx. But consider the extension of 'F', the set of objects satisfying 'F'. It contains 0 and is obviously hereditary.

Now x is a natural number; that is, it is in all such hereditary sets containing 0. So x is also in the extension of 'F', and thus Fx, qed. The proof, however, depends on the existence, for each predicate, of its extension—a highly infinitistic assumption.

Quine, instead, inverts the definition of "natural number." For him, x is a natural number if all classes containing *it* which are *backwards* hereditary contain *zero*. By "backwards hereditary" (my phrase), we mean that the predecessor of a member is also a member. From this the principle of induction still follows. Assume again that $F0$ and that for all y, if Fy then FSy. Given further that x is a natural number, we prove again that Fx, as follows:

Suppose the reverse, that $-Fx$. Consider the class of numbers in the interval $[0, 1, 2, \ldots, x]$ *not* satisfying 'F'; call this class, 'C'. By our reverse assumption, x is in this class. Now the interval $[0, 1, 2, \ldots, x]$ (written for short '$[0, x]$') is of course backwards hereditary. And from the assumption that if Fy then FSy it follows that if Sy does *not* satisfy 'F', then neither does y. The previous two sentences demonstrate jointly that the class C is backwards hereditary, and we have assumed (by reversal) that x is in C. But since x is a natural number, any class containing it which is backwards hereditary must also, by definition, contain 0—so 0 is also in C. This means that $-F0$, contradiction. The assumption that x, a natural number, might not satisfy 'F' is impossible; so for all x, natural numbers, Fx; qed.

The proof depends on the existence of the class of numbers less than or equal to x which do not satisfy 'F', a subset of the interval $[0, x]$. We need no infinite sets; just infinitely many such intervals. These are provided by Quine's 13.11 in STL. But how are the intervals defined? Clearly $[0, x]$ is the set I_x of numbers z that are less than or equal to x. And this set is definable finitistically à la Quine: z is in I_x if and only if every backwards hereditary class containing x contains also z. Yet this makes I_x transparently impredicative—for I_x is itself a backwards hereditary class containing x. To decide, then, whether z is in I_x we must know first whether z is in I_x.

Thus removal of the Axiom of Infinity does not make the logicist reduction of arithmetic to set theory any less impredicative. (I disagree with Parsons, 1965, p. 197, who states that "for the development of elementary number theory we do not have to suppose that any impredicative classes exist." The disagreement is partly verbal, since Parsons has indicated to me that he had a different notion of "impredicative" in mind.) We still allow the impredicative definition of $[0, n]$—and I am convinced that only our numerical model would allow us to breathe easy with that definition. Similarly, in order to prove an instance of the axiom schema of induction, even in Quine's version of number theory, one must at least assume the existence of a set whose members are those elements z of the intervals $[0, n]$ such that $-Fz$—for any predicate one imagines for 'F', however nonconstructive, and however many alternations of quantifiers needed to express the intended property (*STL*, pp. 74–77). I recognize, of course, that these same properties are countenanced in first-order arithmetic as well; for we must agree that

$$F0 \mathbin{\&} [Fy \supset FSy] \supset (x)Fx$$

for any open sentence imagined for 'F'. But at least the nonconstructivity is here only hypothetical: we claim only that *if* 0 is F, and *if Fy* only if *FSy, then* $(x)Fx$; and to know this, or so it seems to me, less of a grip on nonconstructive ideas is presupposed than in knowing a proposition in which nonconstructive ideas are categorically propounded. So when we *prove* an instance of the axiom schema, in set theory, we use a principle much shakier than what we are proving. For the proof assumes *categorically* the existence of a set defined in terms of the (possibly) nonconstructive predicate imagined for 'F' in the schema (it is the set of members z of $[0, n]$ such that $-Fz$—in Quine's proof).

Certainly, there is a sense in which, given the model in number theory of a fragment of set theory, we *should not* have worried about the fragment—even though we *might* have, if we did not know number theory (Benacerraf, in a communication). It is the sense in which a man might be said to do evil objectively, though relying on all the evidence it was humanly possible to bring together. For the pragmatist, however, this point of view is remarkably metaphysical. For a person *ought* really to follow the best evidence available to him, so that different people in the same circumstances ought to do different things. Once we assume that a mathematician does not know any number theory, that person really ought to worry, because he might have reason to suspect an impredicative theory of inconsistency. The degree of certainty he can claim for his knowledge even of the fragment of set theory under discussion, drops. We thus have a clear sense in which even weakened set theory cannot do the work of number theory.

We have not come yet to the "last efforts of the logisticians." If even the modicum of set theory one needs in developing number theory (surrogates) is more suspect than number theory, why not consider the Peano postulates themselves, in their set-theoretical version, as axioms of class truth in their own right? Epistemically, the "arithmetic" of ϕ, $\{\phi\}$, $\{\{\phi\}\}$, . . . should be no more mysterious than the arithmetic of 0, 1, 2, 3, ...

More specifically, we take the Peano postulates, and "relativize" them, their quantifiers, to a newly introduced *primitive* predicate, say 'Z'. The relativized axiom schema of induction would look like this:

$$\mathbf{A}[0] \,\&\, (y)(Zy \supset (\mathbf{A}[y] \supset \mathbf{A}[Sy])) \supset (x)(Zx \supset \mathbf{A}[x]),$$

or, abbreviated:

$$\mathbf{A}[0] \,\&\, (y)_Z(\mathbf{A}[y] \supset \mathbf{A}[Sy]) \supset (x)_Z\mathbf{A}[x],$$

where 0 is the null set, ϕ, and Sx is to be equivalent with (for example) $\{x\}$. (In other words we might choose the *Zermelo* numbers.) And the other axioms would enunciate truths by

Logicism Reconsidered

legislation, to the effect that each set that is a Z has a predecessor with respect to S, except 0—ϕ, that is; that each member of Z has its successor in Z; and so on, relativizing all the Peano postulates to the predicate 'Z'. (Where I speak of Z as a class, the reader may understand this in the sense of a "virtual" class in the sense of Quine, *STL*, sec. 2, or, if the set theory in question has an axiom of infinity, a real class.) To these relativized axioms, we add the general set axioms, enough to provide full set theory—and presumably we will have to add most of the axioms already familiar to us, so that the completed theory will be little less than ordinary set theory with the *addition* of a certain number of relativized, "number theoretic" axioms. Most prominent among the relativized axioms, again, will be the axiom schema of induction for 'Z', which thus becomes, as it was for Poincaré, unreduced and perhaps a priori.

In a way this vindicates Quine. "Arithmetic" is done solely in a universe of sets, a monistic abstract world. But it is hard to believe that Quine could be happy with this solution, that he had this in mind. Our set theory is not what it was: it contains new axioms, a new kind of axiom. It also contains a new primitive: the predicate 'Z'. We are "reducing" arithmetic, then, to something other than classical set theory in order to achieve ontic unification by force. Some remarks and objections to this course are in order.

First, since our new set theory is no less potent than the old, we can, as before, introduce the *defined* predicate, 'NN', by putting x in NN if *zero* is in every backwards hereditary set containing x. From this definition, the following induction schema follows:

$$(\mathbf{A}[0] \ \& \ (y)_{NN}(\mathbf{A}[y] \supset \mathbf{A}[Sy])) \supset (x)_{NN}\mathbf{A}[x].$$

So do the other Peano postulates, relativized this time to 'NN'. We thus have a dual system of arithmetic, that of 'Z' and that of 'NN'.

Mathematical Knowledge

Now we can show that the two predicates are coextensive, that the Z-numbers are none other than the NN-numbers. Both "sequences," after all, begin with ϕ—the null set; the successor "operation" is the same for both. And both Z-arithmetic and NN-arithmetic, of course, sustain their own principles of induction. So we have that 0 is Z; we know, too, that the successor of a Z is also a Z—thus by NN-*induction*, all NN-numbers are Z-numbers too. Similarly, by Z-induction, it follows that all Z-numbers are also NN-numbers. This proof is possible, however, only after we assume Z-induction as primitive, for Z-induction is independent of the rest of the axioms of set theory—unlike NN-induction.

Since Z-arithmetic is independent of NN-arithmetic epistemically (the two can be shown identical only after Z-arithmetic is accepted, especially Z-induction) it stands on its own feet, acquiring no added plausibility from the Fregean reduction of number theory to the theory of 'NN'. On the other hand, the plausibility of NN-arithmetic rests with the legitimacy of the definition of NN numbers.

Our new set theory thus contains an embarrassment of riches. We have two ways of "reducing" in it number theory to sets. The reduction to NN-arithmetic purports to be a true reduction of number concepts and principles to set concepts and principles. Our objection to it has been the epistemic "risks" inherent in impredicative definitions, the heart of Cantorian set theory. The other "reduction," the reduction to Z-arithmetic, is indeed an ontic "elimination" of numbers in favor of sets; yet to make it succeed we have had to puff up set theory epistemologically by artificially attaching to it the relativized postulates of another discipline.

One might reply that "puffing up" of set theory would take place in any event, since even without the Z-reduction, we would have to adjoin the postulates of arithmetic to those of set theory in order to conduct elementary mathematics. And all things being

Logicism Reconsidered

equal, why not incorporate these postulates in a set theoretical setting? The answer is simply this: What good does such incorporation do? The ontologist, looking in, may rejoice in a diminution of the universe of abstract entities. But to the mathematician himself, the reduction of arithmetic to Z-arithmetic brings no gains, and some trouble.

For, consider, one final time, the situation. The only behavioral difference we will detect in a mathematician who has agreed to "reduce" arithmetic to Z-arithmetic within set theory, is a willingness on his part to entertain and answer questions like "Is 2 a member of 3 or a subset?"—questions which Quine himself calls "don't-cares" (*WO*, p. 182) and which would make no difference to knowledge no matter how answered. Since 'Z' to be safe must be regarded as primitive, unlike the impredicative 'NN', the mathematician does not even obtain the benefit of an explicit definition of the concept, "natural number." It is true, of course, that the mathematician's new-found ability to answer questions like, "Is five in the power set of six?" closes certain "truth-value gaps." Before the reduction such questions had no answers and might have been regarded as meaningless. Now, any question that can be formulated syntactically has a meaning and truth value (barring such persistent problems as the Continuum Hypothesis and other "undecidable" conjectures that, some might argue, will never be granted a truth value). But note that a much simpler way to close the truth-value gaps would have been to declare all the cross-theoretical statements (such as "2 is a member of 3") *false*, instead of meaningless. In fact this suggestion is actually made by Quine in another context (*WO*, p. 229). So the reduction of arithmetic to Z-arithmetic cannot be justified by pointing to the closure of truth value "gaps." And thus it cannot be justified at all. For what masks itself as a reduction in ontology (and so it is, to the outsider looking in) is really the addition of an infinite number of useless theorems to one's stock. This is what the "reduction" amounts to for the working mathematician.

Mathematical Knowledge

The reductionist might object that the "don't cares," the "useless theorems," exist anyhow. After all, $\{\phi\}$ was always a member of $\{\{\phi\}\}$; what is new is the labeling of the former entity "one," and of the latter, "zero." The "fact" that 1 is a member of 2 is thus nothing new.

This objection, however, will not do. We have already demonstrated that, in order to deal with Wittgenstein's attacks on logicism, the reductionist must regard definitions such as '$1 = \{\phi\}$' as part of the object language. These definitions, therefore, even if they are a priori truths, are nevertheless "new."

The conclusions of our study of *ontological* logicism may be briefly stated. There are two ways to reduce the world of numbers to that of sets—either through an explicit but impredicative definition (NN), or through postulation of set-theoretical versions of the Peano postulates (Z). The former approach does purport to reduce the *theory* of numbers to the *theory* of sets but is unjustified because it involves us in epistemic risk (greater risk, that is, than there was in number theory before the reduction) since set theory is shakier than number theory. The latter version is not risky, just otiose. It involves the addition to our theory of infinitely many useless propositions. (Tharp, 1971, p. 163, argues similarly. Though the preceding remarks were conceived independently, they constitute, I hope, an amplification of his ideas— especially his remark, "merely adding new primitive predicates to set theory and doing arithmetic relative to them would not have been taken seriously.")

One might still argue in favor of the explicit definition that we are stuck with set theory anyhow and so that we lose nothing in "total certainty" by this reduction (Benacerraf, in a communication). It is of course true that set theory is indispensable for many of the concerns of mathematics (though this might be contested by certain mathematicians), in fact set theory is needed even in the study of numbers. But the overwhelming majority of arithmetic truths do not require set theory for their verification.

Logicism Reconsidered

By keeping arithmetic, unreduced, together with set theory, we do not increase certainty in the areas of mathematics or science that depend on the aid of set theory. But we do increase certainty where arithmetic is the only mathematical theory we need. Even when arithmetic must be used with some other science to obtain knowledge, it is important to segregate error-prone from error-free theories, in order best to know where to search for revision, should revisions become necessary. In short, whenever we need arithmetical facts, either for their own sake, or for some ulterior motive, the counsel of rationality is to use undiluted arithmetic where undiluted arithmetic alone will do. The notion of "total certainty" of our beliefs is useful only if we are constrained to rest any of our beliefs on all of them. But this is not so, arithmetic need not rest, by and large, on any other mathematical science, and therefore it ought not. Why settle in arithmetic for the certainty that set theory can supply, just because we must thus settle in analysis or ordinal number theory? Even a holist like Quine, if he is also a pragmatist, ought to face reality. It is not I who "fragment the enterprise of knowledge" (Benacerraf in a letter); we find it so.

Nor does Quine disagree with this picture of knowledge. Quite the contrary: "Let us keep in mind . . . that knowledge normally develops in a multiplicity of theories, each with its limited utility and each, unless it harbors more danger than utility, with its internal consistency. These theories overlap very considerably, in their so-called logical laws and in much else, but that they add up to an integrated and consistent whole is only a worthy ideal and happily not a prerequisite of scientific progress" (*WO*, p. 251). Quine soon turns this doctrine, culled from Conant, on the question of nominalism, the existence of classes:

Scope still remains for nominalism, however, and for various intermediate grades of abnegation of abstract objects, when with Conant we think of science not as one evolving world view but as a multiplicity of working theories. . . . The nominalist can realize his predilection in special branches, and point with pride to a theoretical improvement of those branches. In the same spirit even the mathematician, realist *ex officio*, is always glad to find that some particular mathematical results that had been thought to depend on functions or classes of numbers, for instance, can be proved anew without appealing to objects other than numbers. It is generally conducive to understanding to keep track of our presuppositions, in point of objects and otherwise, project by project; and to welcome ontological economy in connection with one project even if a more lavish ontology is needed for the next.   [*WO*, p. 270]

I certainly agree with these sentiments, though Quine is vague concerning "lavishness." One can be epistemologically lavish, in accepting doubtful entities, like ordered pairs; one can be lavish in the ontic sense, in accepting unnecessary entities. Reduction usually cuts down on both forms of "lavishness." Only here, the reduction of arithmetic to set theory, do we hear that we can achieve ontic economy through epistemic largesse. Since Quine does, however, advocate treating separable parts of science separately (where necessary) he ought to agree in refusing to acquiesce in *this* reduction to set theory, while accepting others. Instead of being happy to hear that sets can "do the work" of numbers, we should be grateful that in arithmetic we can largely avoid depending on sets.

One might wonder whether reduction is then ever possible, since all reductions seem to reduce a weak but more certain theory (Boyle's and Charles' laws) to a stronger but less certain theory (molecular theory). Such an objection would overlook a cardinal difference: bona-fide reductions effect changes that improve the original theories. They explain why the originals fail to be universally true. Nor is it true that all reductions replace more certain with less certain theories. The reduction of ordered pairs to set theory, the arithmetization of the calculus, are good precisely because the original theories come out with heightened legitimacy.

Logicism Reconsidered

II

Logicism is found wanting, even Quine's ontological version. There is no *reason* to "get rid of" numbers in favor of classes—even if the feat is not, as Poincaré and Wittgenstein thought, impossible. Previously we found the epistemological, classical, versions of logicism, which ground mathematical in logical knowledge, even more suspect, owing to the infirmities of set theory. This suggests a theme that characterizes the rest of this book: arithmetic as an autonomous science, needing no foundation, whether epistemic or ontological, in some other branch of inquiry. This is a view to which Poincaré, but not Wittgenstein, inclined. The latter referred scornfully to the notion of arithmetic as the "natural history" of the natural numbers and refused to regard arithmetic as a science on a par with zoology. Yet this is precisely the view defended (or at least raised for consideration) in what follows as a serious option for the philosophy of mathematics.

The view that mathematics is a separate science does imply that the natural numbers, the subject matter of the science, are objects in the same sense that molecules are objects. (Here I agree totally with Wittgenstein.) Before we may regard mathematics as a science, therefore, we must examine the arguments of Paul Benacerraf, for his view that numbers are not objects at all spells doom for the purported "subject matter" of our science, and thus for the science itself.

The gist of Benacerraf's ideas on this subject can be gathered from this quotation from his well-known article, "What Numbers Could Not Be":

Any system of objects, whether sets or not, that forms a recursive progression must be adequate. But this is odd, for any recursive set can be arranged in a recursive progression. So what matters, really, is not any condition on the *objects* (that is, on the set) but rather a condition on the relation under which they form a progression. . . . That any recursive sequence whatever would do suggests that what is important is not the individuality of each element but the structure which they jointly exhibit. . . . I therefore argue . . . that numbers could not be objects at all; for there is no more reason to identify any individual number with any one particular object than with any other (not already known to be a number). . . .

The properties of being numbers which do not stem from the relations they bear to one another in virtue of being arranged in a progression are of no consequence whatsoever. But it would be only these properties that would single a number as this object or that. [pp. 69–70]

This view can (and will) be discussed on its merits, but what is surprising is that Benacerraf should hold it at all. For in the same article, further back, Benacerraf espouses a philosophy that contradicts what we have quoted: "What constitutes an entity is category or theory-dependent. . . . One might agree with Frege, that identity is unambiguous, always meaning sameness of object, but that (contra-Frege now) the notion of an *object* varies from theory to theory, category to category—and therefore that his mistake lay in failure to realize this fact" (p. 66).

But if the notion of 'object' is relative to category or theory, how can one possibly conclude that numbers are not objects (as Benacerraf does) by showing that numbers are not identical to any given progression of objects outside the category of numbers? (Benacerraf holds of course that sets are not part of number theory; otherwise Frege's "mistake" would not illustrate this point of the senselessness of cross-category identities.) To say that the "properties of number which do not stem from the relations they bear to one another in virtue of being arranged in a progression are of no consequence whatsoever" is significant only against the *Fregean* view of objecthood. Against such a view it might be relevant to conclude that numbers are not objects, from the fact that number theory does not include enough to "flesh out" the numbers, enough to pick them out from the entire universe. But given Benacerraf's own views on objecthood, according to which identity statements are "semantical" only

Logicism Reconsidered

when the terms belong to the same theory, one should say that *no* system of objects, other than the numbers themselves, can even in principle be identified with the numbers, and that numbers are not only objects, but irreducibly so.

Incidentally, Benacerraf makes sport of Frege for believing that such sentences as "Two equals Julius Caesar" have truth values. But as Parsons points out (1965, pp. 188–189) a deeper reason that Frege had for identifying numbers with particular objects was to give a semantical analysis for sentences of the form

$$N_xFx = y$$

with 'y' a *variable*. In this his move is similar to Quine's (*WO*, p. 258) with respect to ordered pairs, which we have discussed earlier.

But ignoring, for the moment, Benacerraf's earlier view on objecthood in general, and adopting for the while the "naive" Fregean view, we still can examine the particular thesis that numbers are not objects. Does the logicist hunt for the "real" natural numbers end in such embarrassment of success as to destroy the object of the search? Benacerraf's conclusion is this:

Therefore, numbers are not objects at all, because in giving the properties (that is, necessary and sufficient) of numbers you merely characterize an *abstract structure*—and the distinction lies in the fact that the "elements" of the structure have no properties other than those relating them to other "elements" of the same structure. If we identify an abstract structure with a system of relations (in intension, of course, or else with the set of all relations in extension isomorphic to a given system of relations), we get arithmetic elaborating the properties of the "less-than" relation, or of all systems of objects (that is, *concrete* structures) exhibiting that abstract structure. That a system of objects exhibits the structure of the integers implies that the elements of that system have some properties not dependent on structure. It must be possible to individuate those objects independently of the role they play in that structure. But this is precisely what cannot be done with the numbers. To *be* the number 3 is no more and no less than to be preceded by 2, 1, and possibly 0, and to be followed by . . . . Any object can *play the role of 3*; that is, any object can be the third element in some progression. What is peculiar to 3 is that it defines that role—not by being a paradigm of any object which plays it, but by representing the relation that any third member of a progression bears to the rest of the progression. [p. 70]

However, the conclusion is obscured by two factors: (1) a heavy load of apparent ontological commitments (what are "elements"? "roles"? "structures"?) such that it is hard to tell which are meant seriously; and (2) a suggestion of a variety of tenable interpretations.

One interpretation of the passage would give a position entertained by Benacerraf in his 1960 thesis. The number 3, for example, would be the class of all third members of progressions. This is the "ordinalist" position, in distinction from the more widely known "cardinalist" position, according to which the number 3 is set of all three-membered sets. Of course, according to this position, numbers emerge again as particular sets, since whatever the significance of choosing one progression rather than another, a progression is nonetheless chosen. (Another problem with this position is that the set theory used by Benacerraf to support his definition of number was discovered to be inconsistent after the appearance of the thesis.)

The second interpretation of the passage would regard all its Platonist overtones as just a manner of speaking. The position would be in fact that all statements of arithmetic are really hypotheticals. Arithmetic is the study of progressions and not any particular progression; arithmetic is the study of models of the Peano postulates, just as group theory is considered the study of all and any models of the group postulates. "Three plus two equals five" means "In any progression, the third, second, and fifth members are related in the manner of addition." Again, "There are 234,456 stars in the galaxy" might be translated, "Every progression of order-type ω is such that the first 234,456 members can be put into correspondence with the stars in the galaxy."

This position does not take arithmetic as it is, but rather does a rewrite. Usually, one regards numbers (whatever they may be) as the building blocks that make up the "concrete" structures that exemplify the group postulates and other undeniably "abstract" structures. If the Benacerraf view (on the present interpretation) were to prevail, something else would have to serve as building blocks—presumably sets. Further, we need a guarantee that there is at least *one* progression of order-type ω for otherwise "there are 234,456 stars in the galaxy," analyzed as above by hypothesizing the existence of such a progression, might come out vacuously true. Sets need to be drafted to make sure of the existence of at least one progression. But how are we to be sure that *these* exist? If we stick at the number 3, because its only properties are to be greater than 0, 1, 2, and less than 4, 5, 6, . . . , what are we to say of the set containing only the empty set? It, too, has no properties other than containing ϕ, and of being a member in various other sets. At least set theory does not capture any other properties. (Though this line of reasoning was thought out independently, Hilary Putnam, 1967, makes a similar argument. He, however, continues to take "if-thenism," as he calls it, seriously, for reasons that appear mysterious to me.)

A third possibility is the wildest, though in some ways the most faithful to what Benacerraf seems to be saying. That is to try to make sense of the notion of an "element" in an abstract structure, where the structure, though not the element, is viewed as an object. The elements of such a structure can be the values of variables (we do not need to rewrite arithmetic), hence Benacerraf's parting shot at the end of the article. I shall leave the reader to make sense of this; I think that where the view goes bad is further back, when Benacerraf presents his evidence for his final position.

Mathematical Knowledge

For even if we accept the Fregean point of view, such that a priori there is nothing wrong with identifying numbers with "something else," from a different category, it does not follow that any such identification must be successful. If our critique of Quine holds good, it shows that the entire view that we have a right to "reinterpret" number theory needs examination. It shows that it is misleading to say "Anything can play the role of 3." For in my view, none of the infinite ways of reducing number theory to set theory is justified.

Benacerraf and Frege both take logicism too *seriously* as a philosophy of mathematics, and hence both overreact to its downfall. Both take too seriously the ambitious claim of logicism to serve as a foundation for mathematics: Frege, the claim of logicism to provide an epistemic foundation; Benacerraf, its claim to provide an ontological foundation. When confronted by Russell's paradox, which undermines logicism's epistemological pretensions, Frege sees arithmetic itself as tottering. And Benacerraf sees, in the failure of logicism to show that numbers are sets, the ontological doom of arithmetic. The view of arithmetic that I proffer for consideration, that of arithmetic as an independent science, avoids both consequences.

Earlier, I noted that Benacerraf's view of objecthood in general upsets his approach to the objecthood of numbers. What I now urge is that, while his view of objecthood might be rejected as general philosophy (on this I take no position), it still could be reconstructed as an a posteriori truth about numbers—that they, at any rate, cannot be interpreted outside number theory. For there is no science more basic than number theory at present available.

Three

Proof and Mathematical Knowledge

The view of mathematics as peculiarly *deductive* is widely shared. Philosophers of the most diverse outlooks agree that the special role of *proof* in mathematics distinguishes this science from all others. In principle, it is held, mathematical proof is essential to mathematical knowledge. This view seems implicit in both logicism and formalism as classically formulated, though perhaps logicism is conceivable without it. The view also has cultural support: it is the one generally taught in the schools. Finally, journals of mathematics will not publish anything less than a proof of a scholarly result.

With the failure of logicism in the preceding chapter, we began toying with the notion of mathematics as an autonomous science whose distinctiveness lies in its subject matter, not in its aims or methods. But the picture of mathematics as a purely deductive science certainly casts suspicion on this naive view of mathematics as the "science of number." For the putative subject matter of arithmetic is, of course, the natural numbers, and why should *they* be open only to knowledge by proof?

Mathematical Knowledge

I propose in this chapter, therefore, to deal with the relation between proof and knowledge in mathematics in utter abstraction from the classical philosophies of Russell or Hilbert. Their specific views will not be discussed, but rather an unspoken presupposition that has dominated almost all of the philosophy of mathematics.

(Since, in the end, I reject the view that proof is essential to mathematic knowledge, I shall allow myself, in setting forth the view, to err on the side of weakening it, in order to allow my antagonist moves that Russell or Hilbert could not have made. In any event I reject even the weakened positions. Should the reader raise an eyebrow at some of these moves, let him remember that they are allowed only as charity toward an opponent, not as bias toward a friend.)

The position that proof is essential to knowledge in mathematics is decomposable into the following triad:

(T1) If someone knows a mathematical proposition, then he knows a proof of the proposition.

(T2) If someone knows a mathematical proposition, then he knows that the proposition is provable.

(T3) If someone knows a mathematical proposition, then the proposition is provable.

Theses one to three are listed in deductive order—each one implies the succeeding ones; refuting any one refutes the preceding ones. One might deny that (T1) implies (T2), that knowing a proof of a theorem implies knowing that the theorem is provable. For example, it might be possible to know of a given proof, that it is the proof of S, but not to know that S is provable; the concept of "provability" might escape one somehow. Or suppose we interpret (T1) as saying that if someone knows that S, then he knows how to give a proof of S when asked *how* he knows that S. (Connoisseurs will detect here a confusion of use and mention, which could be avoided by circumlocution.) He might then not know of the proof that it is a *proof*, even though he gave it as his reason for believing *S*. Though he knew a proof of *S*, he did not know *S* to be provable. There are many subtleties possible. For our purposes, however, (T1) asserts both that the person can give a proof and that he knows that it is a proof. In the real world, these go together: anyone sophisticated enough to give a proof in complex variables is sophisticated enough to know that he can. Our main task, however, will be to examine what it is to know a proof.

One may wonder whether (T2) implies (T1), whether one can know that a proposition is provable without knowing the proof. For one thing, one can know that a formula is a theorem of a propositional calculus without knowing the proof: one need merely know that the formula is valid, a tautology. But perhaps the argument from the completeness theorem is proof enough, perhaps proofs need not be limited to object-language proofs. But even in the higher realms of mathematics, where completeness is not attainable, one often knows that certain theorems are provable without being able to give the actual proof. One gets an insight into the method of proof without yet having worked out the tricky details. One might, of course, have been wrong about the insight (maybe Fermat was, in his famous *marginalis*), but, as Moore pointed out often, certainty is perfectly compatible with *possible* (in the modal sense) error. That one errs does not imply that one *never* knows.

What is the importance to epistemology of (T1–3)? The answer depends upon *why*, as much as whether, they are true. Consider (T1). It might be that for some trivial reason one always knows a proof of a proposition if one knows the proposition. Define proof from premises as in Church or Mendelson, and suppose that all proof is formal proof, from such premises as are known. Result: when one knows *S*, he can give a proof—the one line consisting of *S*—which is a proof of *S* from a known premise. We might as well then forget (T1). For (T1) to be important, the evidence that (T1) is true must show also that the postulated proof is *responsible* for the knowledge in question. It would also be nice to know whether any other grounds for mathematical knowledge than proof are even possible, given (T1) (the modal interpretation of (T1)). Since I later reject (T1) in all forms, I shall not press the point.

For (T2) and (T3), the question of relevance is sharper. The condition for knowledge proposed by (T3) is not sufficient, but at best necessary, for mathematical knowledge. Even if (T3) be true, we are still left without a theory of knowledge in mathematics. (T3) is compatible even with Platonic mysticism, according to which one truly knows mathematical truths by a process of "apprehending" mathematical objects. (T3) would insist only that this ability is contingent upon the provability of what is learned through apprehension.

(T1–3) are formulated to hold of mathematical knowledge in general. We could alternatively hold that mathematical proof is involved in only *some*, even most, mathematical knowledge. Rewriting (T1–3) to suit, we call the result (T1'–3'). One who held (T2'), for example, might agree that it is not necessary to know, of all the truths that he knows, that they are provable. He might believe that mathematical intuition apprises him of the truth of the axioms of set theory, but where intuition fails, one can know truth only through knowledge of provability. He might go on to say that the Continuum Hypothesis is unknowable, since it is undecidable by proper proof and since intuition gives no verdict.

We have delayed specifying what "proofs" are and what "knowing a proof" is. The following section is an effort to put things right.

II

Proof is formal proof. Arbitrarily, we pick a system—Church's "applied first-order functional calculus." Then, proof is proof from premises (including the null set of premises) in Church's sense. Usual usage is looser: (1) because higher order systems might be used, and (2) because informal arguments are universally described as proofs. In answer, (1) for our purposes now, it is not necessary to decide whether first-order logic alone is logic; and (2) the mathematical community, almost without exception, has been persuaded that no proof is rigorous if not "formalizable." The success of the *Begriffschrift* and its successors in reconstructing mathematical reasoning has given these logistic systems the status of standards, against which informal arguments are to be tested. But once one agrees that nothing is a proof if not "formalizable," there is little interest surrounding exactly which object should be labeled "the proof." If one chooses, as I do, to style the formal proof only, "the" proof (the Platonic ideal in virtue of which the informal argument is valid), one must correspondingly loosen his criteria for *knowing* a proof, to make sure that people can know algebra even if they do not know Church.

Many formal proofs (I shall henceforth leave out the redundant "formal") do not require any but logical premises: they are proofs from no premises, from the null set of premises. By logical premises, I mean the Churchian axioms for the first-order functional calculus. For example, if we write all the theorems of group theory in the form, "If G is a group . . . ," the group axioms thus absorbed into the antecedents, all such theorems are logically true, valid. (Better, they are valid if their predicates are regarded as uninterpreted, or as variables in the way of Church, p. 324.)

Some, most, mathematical theorems are categorical: they cannot be proved without the use of nonlogical premises. But what premises are allowed in the proofs required by (T1–3)? Certainly not 'p & $-p$' or other false or inconsistent premises. Even true premises will not necessarily do, for then (T1–3) will lose significance by being trivially verified; any known (thus true) proposition will have as proof the one-liner consisting of the proposition itself. For the same reason, it will not be enough to restrict premises to *known* truths; again (T1–3) would receive trivial verification.

If (T1–3) are to retain interest, their proponent will have to rely solely upon *standard* premises, independently (of the particular theorem in question) accepted by the mathematical community as usable in proof without any further need of argument (Compare Shoenfield, pp. 1–2). Standard premises need not constitute an unchanging totality; their identity and number might change from time to time. What is standard (legitimately) for one generation might not be standard for another. However, standard premises ought to be community-wide at any given time.

Gödel's first theorem prevents us from identifying the "standard premises" with any of the current axiomatizations of arithmetic or analysis. And if we support the view that, as we add Gödel sentences to the axioms of arithmetic, we continue to generate new knowable Gödel sentences, we would have to conclude that the set of standard premises is nonrecursive. No recursive extension of the arithmetic axioms would suffice to yield *all* the possible Gödel sentences obtained by Gödelizing these extensions. But to hold that the standard premises are, at any moment, a nonrecursive totality is distasteful. One could, however, flatly deny that *all* Gödel sentences are knowable—at some point, they become too much for us (suggested by Richard Grandy). If so, the standard premises so far could even be finite in number. On the other hand, it seems that all of the current axioms are standard premises, for the axioms are actually used in securing knowledge. Certain major theorems are also standard, for though they flow from the axioms, they may be more certain than the latter.

I shall say no more about standard premises. My remarks are designed, not to defend the notion, but rather to give the reader some sense of the moves the proponent of (T1–3) must make: he must grant the existence of standard premises, but his characterization of them can be weak. They could vary over time, and they need not even at any time be a recursive totality.

One problem is pressing. That is, do (T1–3) apply to the standard premises themselves? We may, of course, limit our theses to nonstandard premises, asserting only what I called (T1'–3') a while back. Otherwise, there remain these options:

(1) (T1–3) are trivially true of the standard premises, since they have *themselves* as one-line proofs. But since standard premises are special, the trivial proof can somehow be considered grounds for believing them. One who looks at proofs as a "game" might hold this view: the rules of the game allow us to write one-line proofs, and there is no distinction in principle between short proofs and long ones.

(2) Standard premises have nontrivial proofs, from other standard premises. This would create an infinite hierarchy of standard premises. But this option seems fantastic; if proof must proceed from *known* premises in order to create knowledge (a reasonable assumption), then to know any one thing, we would have to know an infinite number. To stop the regress, in the name of (T1–3), we can say either

(a) that at some point all proof is circular, a view associated with Hegel in the realm of general knowledge, or

(b) that at some point, the standard premises cease to be mathematical, so that (T1–3) cease to be relevant.

We now turn to the concept of *knowing a proof.*

III

One certainly knows how to prove something, if one can write up its formal proof from premises in a first-order logic, say Church's, using only standard premises. As a *necessary* condition for knowing a proof, however, this will not do, for it follows, from the condition, that most mathematicians have never known any proofs. What we need to say is that a mathematician is in principle able to give a formal proof, and hence knows the proof, if and only if the only obstacle is his lack of familiarity with formal logic and Church's system in particular. (It is helpful to choose a particular system to fix our ideas.) A mathematician is said to know a proof if he could, if practiced in logic, write it up formally, using only standard premises. This still misses the mark, for some mathematicians might be mediocre logicians after a lifetime of practice. The art of formalizing proofs may demand an obsessive-compulsive character, a character incapable of mathematical creativity.

A device of Socrates' will help here. Let us assign a logician as a "midwife" to our mathematician. The logician will supply no premises; he merely transcribes the informal arguments of the mathematician into Church's calculus. If the two together can bang out a formal proof, then the mathematician is said to have known the proof all along, on the basis of the informal argument. So a mathematician is said to know a proof of S, if, working with a logician who supplies no premises, he could produce a formal proof of P (i.e., the *wff* which expresses S) in Church's system from standard premises only.

Imagine the steps occurring in the collaboration between the two. When asked, "How do you know the theorem S?", the mathematician will reply by giving an informal argument. The logician, we assume, knows no mathematics, and thus cannot yet follow. The mathematician has skipped many logical steps, as is his wont, and he has not stated all his premises. Indeed, he is probably oblivious to the fact of his omissions. The logician-midwife must therefore draw the mathematician out, forcing the latter to be explicit. This is a delicate operation, for under persistent questioning, the mathematician may discover, and plug, real gaps in the proof. On the spot he may discover important premises as yet unproved and prove them. The collaborators will have to be careful to make explicit only those premises and arguments that were implicit in the mathematician's initial response.

Assume, then, that the mathematician, upon being prodded by his midwife, brings to light all the premises he had in mind, and they are standard. We have now a supplemented informal argument that expresses no more knowledge than the mathematician had before. The logician must now take over and produce from this a formal proof.

First, the logician must translate all of the supplemented informal argument into sentences of the first-order calculus. In doing so, he may use whatever predicates he needs, introducing them into the (applied) functional calculus. If he cannot do the translation, we say that the informal argument is incomplete, and hence that the mathematician did not know a proof of 'S'.

But suppose the translation successfully completed. The result is a sequence of well-formed formulas that are the translations into first-order logic of (1) the standard premises the mathematician used; (2) the conclusion; (3) the steps of the argument. This sequence almost certainly will not be a formal proof; to tell, therefore, whether the mathematician knew a proof of the conclusion, we must see whether the logician can supplement this sequence with other well-formed formulas so that the result is a proof, from the premises of the conclusion. The logician, remember, may supply no further premises. We have two cases:

(1) If the logician cannot come up with a formal proof, this is perhaps because the standard premises mentioned simply do not imply the conclusion. But suppose, now, that the standard premises supplied do imply the conclusion, but that the logician fails to discover a completed proof of this fact. The failure could be blamed either on the mathematician or the logician. The crux of the matter is the kind of logician we envision. If he is dull, his failure should not be laid at the doorstep of the mathematician's alleged ignorance. On the other hand, we cannot envision a

superhuman, because such a being would discover a completed proof despite the ignorance of the mathematician. What we need, then, is a mind that is brilliant at analysis and symbolic manipulation, a logician who could with ease turn the mathematician's contribution (if it is sufficient) into a formal proof, but who lacks mathematical creativity. The distinction I make between the two skills is real, even if the boundary between them is vague. We find surprising agreement among professionals in ascribing these skills to people. If *such* a logician fails to find a proof, we ascribe ignorance to the mathematician despite his informal argument and despite the formalizability of his argument by a logician of *mathematical* talent.

(2) If *such* a logician (one lacking in mathematical creativity, but versed in logical manipulations) does manage to produce a completed formal proof, we credit the mathematician with having known the theorem all along, on the basis of his informal argument.

The reader will complain and we will readily admit that the above test fails to distinguish knowledge from ignorance on the part of the mathematician in ever so many cases. In practice, we often should not be able to tell whether a mathematician really knew a theorem or whether the logician was instrumental (rather than being just a catalyst). This uncertainty may convince the reader that the concept of 'knowing a proof' is vague itself or, alternatively, that our analysis of the concept is lacking. But our results are precise enough for the use to which we shall put them in the next section. Our test cases will involve no such uncertainty, and with these cases we hope to undermine (T1–3).

IV

I proceed, then, to show that (T1) and (T2), in particular, are false, by presenting a case of knowledge without proof and without even knowledge of a proof's existence. This is not to say that knowledge is possible without "evidence." But the evidence in my example is inductive, not deductive, and the example is taken from G. Polya's book, *Induction and Analogy in Mathematics*, section 2.6. Almost no one, including the author of the book, has ever noticed (in print, anyhow) that *some* of Polya's illustrations exemplify not only how mathematical truth is discovered, but also how it might be *known*. This insight is the only one I wish to claim, in this section, as my own.

Bernoulli set the mathematical world the problem of finding the sum of the reciprocals of the squares

$$1 + 1/4 + 1/9 + 1/16 + 1/25 + \ldots$$

Euler found the solution by the following reasoning: if an equation of degree $2n$ has the form:

$$b_0 - b_1 x^2 + b_2 x^4 - \ldots + (-1)^n b_n x^{2n} = 0$$

and $2n$ different roots:

$$r_1, -r_1, r_2, -r_2, \ldots, r_n, -r_n$$

then

$$b_0 - b_1 x^2 + b_2 x^4 - \ldots + (-1)^n b_n x^{2n}$$

equals

$$b_0(1 - x^2/r_1{}^2)(1 - x^2/r_2{}^2) \ldots (1 - x^2/r_n{}^2)$$

and

(A) $$b_1 = b_0(1/r_1{}^2 + 1/r_2{}^2 + \ldots + 1/r_n{}^2).$$

If we consider the equation

$$\sin(x) = 0$$

we can expand the left side, resulting in

$$x/1 - x^3/3! + x^5/5! - x^7/7! + \ldots = 0.$$

Suppose we regard this as an equation of "infinite degree" (here is where the inductive "leap" comes, unjustified by anything so far presented) whose "roots," the zeroes of $\sin(x)$, are, of course,

$$0, \pi, -\pi, 2\pi, -2\pi, 3\pi, -3\pi, \ldots$$

Dividing by x, to remove the root 0, we obtain the equation

$$1 - x^2/3! + x^4/5! - x^6/7! + \ldots = 0$$

with the roots

$$\pi, -\pi, 2\pi, -2\pi, 3\pi, -3\pi, \ldots$$

By analogy with (A), we conclude that

$$\begin{aligned}
\sin(x)/x &= 1 - x^2/3! + x^4/5! - x^6/7! + \ldots \\
&= (1 - x^2/\pi^2)(1 - x^2/4\pi^2)(1 - x^2/9\pi^2) \ldots
\end{aligned}$$

and

$$1/3! = 1/\pi^2 + 1/4\pi^2 + 1/9\pi^2 + \ldots$$

so that, finally,

(B) $$1 + 1/4 + 1/9 + 1/16 + \ldots = \pi^2/6.$$

This was Euler's discovery, the solution to Bernoulli's puzzle.

But what grounds did Euler have—what grounds could he have had—for believing (B)? Euler had methods of approximating π, and he calculated the value of the series

$$1 + 1/4 + 1/9 + \ldots$$

to many places, finding that the value coincided with the places of $\pi^2/6$. True, it is generally held that because a mathematical proposition is verified for $n = 1$ to $n = 20$ one cannot conclude that it holds generally. But Euler's results should give pause to such dogmatism. What we know, what *he* knew, about analysis made it impossible to believe that 20 places of the series

$$1 + 1/4 + 1/9 + 1/16 + \ldots$$

should coincide with $\pi^2/6$ by "accident." The value $\pi^2/6$, after all, was not concocted ad hoc, as an approximation—it fell out of the blue, the outcome of operations that had initial plausibility.

Proof and Mathematical Knowledge

Further, he knew that a constant like π, on the basis of past experience, was likely to show up in such a context. Finally, there was no competing value available, while the available value, $\pi^2/6$, was simple and esthetic. Under the circumstances, a knowledge claim was almost inevitable, and almost justified.

But there was yet more evidence, which in my opinion absolutely settled the question as to whether $\pi^2/6$ was truly the sum. First, Euler expressed other coefficients of the expansion of $\sin(x)$ using infinite sums. He achieved the result that the sum of the reciprocals of the fourth powers,

$$1 + 1/16 + 1/81 + 1/256 + 1/625 + \ldots$$

equals $\pi^4/90$, a result that again he verified to many decimal places. Second, he considered the equation

$$1 - \sin(x) = 0$$

with the roots

$$\pi/2, \pi/2, -3\pi/2, -3\pi/2, \ldots$$

(The curve is tangent to the x-axis at infinitely many points.) By similar decomposing, he arrived at the equation:

$$\pi/4 = 1 - 1/3 + 1/5 - 1/7 + \ldots \text{ (See Polya for details.)}$$

But this result had actually been demonstrated by Leibniz! Here was a major breakthrough: Euler's new analytic method yielded results that not only checked, but that were demonstrable. Polya discusses many other ways that were or could have been used to test Euler's method by deriving new truths or old ones already verified. The weight of the evidence was such that no rational being could deny at least this: that $\pi^2/6$ is the sum of the reciprocals of the squares,

$$1 + 1/4 + 1/9 + \ldots,$$

nor could anyone today after reading the last few pages. Euler could not legitimately doubt his result, and he is justly celebrated for having discovered it.

This is not to say that Euler had *proved* the result. The reader can easily convince himself by rereading our remarks in the preceding section that Euler was very far from knowing a proof of his solution to Bernoulli's problem. Further, Euler (as Polya points out) knew full well that his argument was no demonstration, even by eighteenth-century standards. No axiomatization, or even formalization, of "infinite addition" existed at the time; this awaited the nineteenth century. (For details, read Grattan-Guiness' fascinating book, *The Development of the Foundations of Mathematics from Euler to Riemann.*) The analogy between finite equations and "infinite equations" was a perilous guess, nothing more. Had Euler attempted to set down all the premises of his argument in mathematical detail and precision, he would undoubtedly have written falsehoods—for the analogy between the finite and the infinite often breaks down, a fact that ultimately created a crisis in mathematics (described in the above-named book). Euler's genius and his painstaking verifications saved him from error in employing the new analysis, in exploiting the analogy between finite and infinite.

Euler had not proved his result: it rested on a method whose principles are very subtle and which Euler did not know. But when we narrow our focus from the method to its product,

$$\pi^2/6 = 1 + 1/4 + 1/9 + \ldots$$

and ask ourselves to what extent *it* was confirmed for Euler, we must admit that Euler had a right to be confident in his discovery beyond any doubt. And so, both (T1) and (T2) are false: (T1), because Euler did not know any proof of his discovery; (T2), because he could hardly have *known* (though he certainly had faith) that a proof of his result, from standard premises, was possible.

Proof and Mathematical Knowledge

The evidence Euler gathered was inductive, of the kind employed in general science. And the true rule of scientific induction in mathematics is yet unsung, though Polya's work has certainly opened avenues of research. Philosophical writers lay it down that no finite number of checks of '$(x)A[x]$' suffice to confirm the generalization, even those who accept such confirmation of generalizations in physics. One never claims, they say, mathematical knowledge on the basis of a finite number of tests, however large. But induction by simple enumeration has been rejected in general science as well; to attack it in mathematics is to attack a straw man. In our present example, 20 decimal places of agreement between $\pi^2/6$ and the sum of the reciprocals of the squares is conclusive evidence that agreement holds throughout. As we have learned from Goodman (1965, pt. III) and others, background information can make a generalization so highly confirmable, that a few positive instances suffice to confirm it. Here, for example, we have the analogy between finite degree equations and infinite degree power series, an analogy that yields other results that are also verified to many decimal places, an analogy that brings forth previously proved theorems. We have the sudden, unexpected appearance of a simple, beautiful result—also considerations in its favor, just as in physics. Not that the analogy at the heart of Euler's discovery is verified thus, except in the trivial sense that we know that it sometimes holds, it is the individual consequence of the analogy, Euler's discovery, which is so verified.

Our purpose here is not to discuss inductive reasoning in mathematics, but to disprove (T1) and (T2). (Many interesting kinds of inductive evidence are sometimes possible in mathematics, but this is not the place to go into them.) We should like to widen the horizons of the reader, hopefully to make him more receptive to the more speculative portions of what is to follow. One of the motives for insisting upon mathematics as a *deductive* science is hostility to Platonism. How ironic, that the extension to mathematics of the epistemic methods of the natural sciences should raise the bogey of Platonism. The fear of Platonism is indeed the primary obstacle in the way of the acceptance of mathematics as a true science. In the next, final, chapter, we must speak to this fear.

Four

Platonism and Mathematical Knowledge

Classical Platonism consists of two distinct doctrines. "Ontological" Platonism should not be confused with "epistemological" Platonism; it is possible to adhere to the former without the latter, or so it would seem.

According to ontological Platonism, the truths of mathematics describe infinitely many real mathematical objects. Since the number of material bodies may very well be finite, most mathematical objects could not be material (compare the beginning of Hilbert's "On the Infinite").

Furthermore, the only way that mathematical statements could be true is by describing such mathematical objects (see Tarski, 1956; Goodman and Quine, 1947; and Benacerraf, 1973). It follows, then, that whether (ontological) Platonism is tenable is the same question as whether the axioms of mathematics are *true*. This conclusion, of course, puts ontological Platonism in a very favorable light. And, following the example of Quine, many philosophers are no longer afraid to call themselves "Platonists."

Epistemological Platonism, however, still raises hackles: this is the doctrine that we come to know facts about mathematical entities through a faculty akin to sense perception, or at least that some people do. As the most distinguished proponent of this view since Plato once put it:

> But, despite their remoteness from sense experience, we do have something like a perception also of the objects of set theory, as is seen from the fact that the axioms force themselves upon us as being true. I don't see any reason why we should have less confidence in this kind of perception, i.e., in mathematical intuition, than in sense perception, which induces us to build up physical theories and to expect that future sense perceptions will agree with them and, moreover, to believe that a question not decidable now has meaning and may be decided in the future. . . .
>
> . . . It by no means follows, however, that . . . because [mathematical intuitions] cannot be associated with actions of certain things upon our sense organs, [they] are something purely subjective, as Kant asserted. Rather they, too, may represent an aspect of objective reality, but as opposed to the sensations, their presence in us may be due to another kind of relationship between ourselves and reality. [Gödel, 1947, in *BP*, pp. 271–272]

But this bold view of Gödel's (to be examined in Section III), by its very effrontery, raises an ancient objection to all forms of Platonism—epistemological as well as ontological.

The objection is that, if mathematical entities really exist, they are unknowable—hence, mathematical truths are unknowable. There cannot be a science treating of objects that make no causal impression on daily affairs. All of our knowledge arises from the causal interaction of the *objects* of this knowledge with our bodies. Since numbers, et al., are outside all causal chains, outside time and space, they are inscrutable. Thus the mathematician faces a dilemma (Benacerraf, 1973): either his axioms are not true (supposing mathematical entities not to exist), or they are unknowable (actually, of course, on either alternative the axioms are unknowable).

Platonism and Mathematical Knowledge

II

The argument, put more carefully, runs thus: it is necessary for knowledge that *what* is known cause the knowledge: the true belief that is to be styled knowledge. Ordinarily, we argue that someone could not have known this or that by showing that he and his sense organs could not have been termini of any causal chain emanating from the thing allegedly known. Further, Platonism concedes, even glories in the fact, that numbers and functions and triangles play no causal role. (According to G. Vlastos, Plato himself held this view.) Hence, the argument continues, if we interpret mathematical propositions in their naive sense (if their truth conditions are as Tarski says) then, it follows, we cannot know any of these propositions. For in the naive interpretation (and what other is there?) the propositions of mathematics refer to these abstracted entities. So goes the argument.

We have here a twin claim: first, that the "causal theory of knowledge" is true; second, that it makes mathematical propositions unknowable. Neither claim can be judged without analysis.

Certainly the causal theory of knowledge cannot be interpreted as

(0) One cannot know that *p* unless *p* causes this knowledge (or belief) that *p*,

for the substitution here of an English sentence for the second occurrence of '*p*' is a blatant violation of the distinction between "use" and "mention." But perhaps the proper phrasing of the matter is this:

(1) One cannot know that *p* unless the fact that *p* causes one's knowledge (or belief) that *p*.

(Of course, this is only part of the causal theory of knowledge, that part which sets out necessary conditions for knowledge.

Some philosophers go further and see in the causal theory of knowledge the total analysis of knowledge—I know that p, therefore, if and only if: (1) p, (2) I believe that p, and (3) the fact that p bears some causal relationship to me. We shall not discuss the causal theory as furnishing *sufficient* conditions for knowledge.)

Here the causal theory of knowledge is set out in all generality, applying to any and all knowledge. Unfortunately, the phrase, 'the fact that p', creates difficulties. For is there, really, any good reason to believe that "facts" exist? The notorious difficulty of formulating a criterion of identity for "facts" is evidence *against* their existence (Quine, *WO*, pp. 246–248). And there is the striking proof by Frege, Gödel, and Davidson (1967a and b) that if facts are "produced" indeed by replacing the letter 'p' in 'the fact that p' by true sentences, such that every such replacement yields a fact-referring singular term, then there is only one fact!

(A version of the argument: If every true sentence corresponds to a fact—*the fact that p*—we have (1) *the fact that p = the fact that $\hat{x}(x = x \,\&\, p) = \hat{x}(x = x)$*, because '$p$' and '$\hat{x}(x = x \,\&\, p) = \hat{x}(x = x)$' are logically equivalent, and we must say that two logically equivalent sentences correspond to the same fact. But, if 'q' also stands for a truth, then $\hat{x}(x = x \,\&\, p) = \hat{x}(x = x \,\&\, q)$, and so we have (2) *the fact that p = the fact that $\hat{x}(x = x \,\&\, q) = \hat{x}(x = x)$*, for we must say that the referent of a referring expression like 'the fact that . . .' depends functionally upon the referents of the referring terms appearing in it. Finally, by the same logical conversion as before (but in reverse), we have (3) *the fact that p = the fact that q*; i.e., if facts correspond to true sentences, then there is only one fact.)

The Davidson-Frege proof does not prove, by itself, that there are no facts or that no singular terms could refer to such things. The argument does purport to show that there is no way to create such a term *from a sentence in general*, for example by saying "the fact that . . . ." But such creation seems necessary if we are even to state the causal theory of knowledge.

What is the trouble? We want the causal theory to apply to any *knowledge that p*, '*p*' to be replaced by any true sentence; yet sentences, and things ostensibly related to sentences, do *not* appear in the logical analysis of causal sentences—as Davidson has argued convincingly in "Causal Relations." What is needed to give the logical form of such sentences is quantification over *events*.

Some philosophers object, however, to the Davidson-Frege proof, balking at one or more of its premises (e.g., Cummins and Gottlieb); insisting that 'the fact that *p*' is a harmless expression—and that facts exist. Can these philosophers hold to the causal theory of knowledge in the form of (1)? No; (1) states that one cannot know that *p* unless a certain fact—the fact that *p*—causes the knowledge. But if facts exist, they certainly are not material bodies. Nor are they events. What else are they, then, but abstract entities, like numbers? If numbers are alleged incapable of influencing the course of events, why should facts be any different? If (1) is true, therefore, *all* knowledge, not only mathematical knowledge, is impossible.

The causal theorist might regroup. In speaking of "causing," in the causal theory, he might say, I did not mean the causal *relation*; my intention was not to say that there is some entity, a fact or a state, that causes anything. What I meant was simply that the sentence replacing '*p*' in 'knowing that *p*' should enter into a causal *explanation* of why it is that the knower knows that *p*. Accordingly, the causal theory of knowledge should be rephrased as:

(2) One cannot know that a sentence *S* is true, unless *S* must be used in a causal explanation of one's knowing (or believing) that *S* is true.

If the causal theorist adopts (2), however, he loses all ability to impugn the Platonist interpretation of mathematics. We can let him assume without question the truth of (2) (let him assume, even, that he knows what a "causal explanation" is) without giving him any weapon against Platonism. For suppose that we believe, for example, the axioms of analysis or of number theory. We can assume that *something* is causally responsible for our belief, and that there exists a theory, actual or possible, known or unknown, which can satisfactorily explain our belief in causal style. This theory, like all others, WILL CONTAIN THE AXIOMS OF NUMBER THEORY AND OF ANALYSIS. Now it is generally accepted that in order to provide an explanation, or to be a member of a set of premises which provide an explanation, a sentence must at least be true. And the Tarskian-Platonist interpretation of mathematical propositions is the only one known to guarantee the truth of the axioms and theorems. Thus, the axioms of analysis, as interpreted by the Platonist, will indeed necessarily be used in whatever causal explanation can be given of our belief that the axioms, as interpreted by the Platonist, are true. But this satisfies (2) perfectly, and the challenge is again turned back.

I conclude that it is impossible even to state the Causal Theory of Knowledge in a way (a) that applies equally to all "knowledge that p," and (b) that implies that mathematical statements are unknowable if true.

The causal theorist is not yet defeated, however, for he can restate his theory in a more restricted way. For example, he might say:

(3) One cannot know anything about F's, unless this knowledge (or belief) is caused by the F's, or some of the F's, etc.

This raises a tangle of problems traditionally associated with the word 'about'. Is the sentence, "There are no F's," *about* F's? This seems a serious problem to me, and I have nothing original to say about it.

Platonism and Mathematical Knowledge

Another problem with (3) is that it imputes causal efficacy to "*F*'s," that is, to all sorts of inanimate objects. Again following Davidson, it is events that are the most likely candidates for "causes" and "caused" alike. True, *conditions* are often said to be causes; the weakness of the bridge caused its collapse. But, as Davidson points out in "Causal Relations," these cases are best subsumed under causal *explanation*, not the causal relation, for precisely the reasons already given above, in our discussion of "facts."

That objects, bodies, cause anything at all seems too animistic to be plausible; it is hard to believe that pieces of paper are *agents*. Still, (3) is not beyond repair. If lampshades do not cause anything, events connected with them do. The collision of photons (an event) with the lampshade may cause within us the belief that there is a lampshade in the room. The causal theorist says that if there really were no such events, we could not know anything about lampshades, to say nothing of photons. The causal theorist, then, views all objects (those, at any rate, which we know anything about) as *participating* in events that cause us to believe (know) about these objects. For example, a person participates in his own birth and in sundry activities. Passively, he participates in collisions of photons with his surfaces, which allow others to know he is around. If we extend further the concept of participation to inanimate things, electrons participate in psi-wave collapses, collisions, and so forth. Generally, the claim is:

(4) One cannot know anything about *F*'s unless this knowledge (belief) is caused by at least one event in which at least one *F* participates.

The causal theorist would argue further that nothing can participate in an event unless it has spatial and temporal position. Numbers and functions, usually thought of as "outside" space and time, are thus unknowable.

But how plausible is (4), really? Suppose a now extinct species of

115

animal roamed the forest, leaving footprints. Zoologists entering the forest see the footprints, i.e., depressions in the soil, and learn about these animals. Now the animals certainly "participated" in events that resulted in the footprints, i.e., in the depressed condition of the soil. And certain regions of the soil participated in events that caused the zoologists to believe in the existence of a hitherto unknown species. But note that the animals themselves did not participate in these later events (unless the notion of 'participation' is distorted beyond recognition). And the events that the animals did participate in certainly did not cause either my beliefs or the events that did.

What happened is that an event led, not to another event, but to a condition, and the condition was partly "responsible" for another event, an event which in turn caused certain beliefs to form in the mind of the zoologist. But surely Davidson is right here: talk of conditions causing or being caused is really talk of causal explanation, and this causal story should be retold in the "formal mode." Once causal explanation is invoked to justify a bit of knowledge, the causal theorist must retreat to version (2)—which is, we have noted, compatible with Platonism and with mathematical knowledge.

There are, remember, other reasons to retreat. Version (4) of the causal theory is anyhow limited to sentences "about" some type of entity. Even if we could say what "aboutness" is, a causal theory limited to an arbitrarily chosen subset of our knowledge is highly suspect. Besides, what is wrong with version (2)? If a piece of knowledge is justified using a causal explanation of it, what more needs to be said? The burden of proof would seem to rest on the proponent of a more restrictive causal theory of knowledge.

In summary: the most plausible version of the causal theory of knowledge admits Platonism, and the version most antagonistic to Platonism is implausible.

Platonism and Mathematical Knowledge

III

The causal theory of knowledge is an extreme version of a less controversial cousin, the causal theory of perception. This theory states, as necessary truth, that perception occurs only if what is perceived causes the perception. The CTP is thus obviously relevant to the truth of Platonism, that version that insists upon a direct "vision" of mathematical truths or of mathematical objects. For again, abstract objects do not, it seems, interact causally with the universe, with our physical surfaces.

Two formulations of the causal theory of perception come to mind, similar to our recent formulations of the causal theory of knowledge:

(A) One cannot see that S ('S' replacing an arbitrary sentence) unless N ('N' replacing a name of the arbitrary sentence) must be part of a causal explanation of the perceptual experience that is part of *seeing that S*.
(B) One cannot see an F, unless the F participates in an event that causes one to have a *perceptual experience of an F*.

If (A) were acceptable, the Platonist could adhere to it, thus accepting a version of the CTP which is entirely compatible with the existence of mathematical "intuition." For if N be an arithmetic truth, certainly it will appear in any explanation of any perception. This argument, however, legitimizes only statements of the form

I see (intuit) that S.

where 'S' stands for an arithmetic truth. And whether such sentences can be considered perceptual is debatable. True, 'I see that you are angry' expresses a perceptual fact; by using my eyes I come to the belief that you are angry through no process of inference, in the everyday sense of that word. What is salient about this example, however, is that besides seeing *that* you are angry, I see *you* (and perhaps your anger, on some accounts of properties). It might be argued that for "seeing that" to be perception, some*thing* appropriate must be seen. The intuition of truth would thus depend for its sanction upon the intuition of objects.

We may not be able, therefore, to escape considering the intuition of abstract objects in the light of (B), the causal theory of perception.

Actually, of course, (B) must be immediately amended, for it is manifest that, in order to see an *F*, one need *not* have the experience of perceiving an *F*. In a graveyard at night, one might have the experience of seeing an octopus, but be *seeing* only a tree. In order to see an *F*, however, one must undergo *some* experience, related in some yet-to-be specified way to the experience of an *F*. Thus we can weaken the causal theory of perception considerably and still threaten Platonism with

(C) One cannot see an *F*, unless the *F* participates in an event that causes one to have *some* perceptual experience.

Nor can we refute (C) by our example of the prehistoric animal, because in our example, the zoologists indeed never *see* the animal.

Why should we believe (C)? A thesis like (C) may be regarded either as stating a scientific hypothesis about our perceptual apparatus, or else as stating an (alleged) necessary condition governing the concept of perception (Grice, 1965, pp. 438–439). We concern ourselves only with (C) in the latter role, as philosophical theory. Grice's argument in favor of (C), or something like (C), is an argument from hallucination. (This argument, of course, establishes the causal theory only as a necessary condition of perception; (C) is worded to reflect that fact. Grice has further arguments and refinements that are intended to establish the CTP also as a sufficient condition of perception. But since we are discussing the causal theory in the context of Platonism, we

Platonism and Mathematical Knowledge

may limit our discussion to (C).) In Grice's classic treatment, the argument emerges as follows:

> Suppose that it looks to X as if there is a clock on the shelf; what more is required for it to be true to say that X sees a clock on the shelf? There must, one might say, actually be a clock on the shelf which is in X's field of view, before X's eyes. But this does not seem to be enough. For it is logically conceivable that there should be some method by which an expert could make it look to X as if there were a clock on the shelf on occasions when the shelf was empty: there might be some apparatus by which X's cortex could be suitably stimulated, or some technique analogous to post-hypnotic suggestion. If such treatment were applied to X on an occasion when there actually was a clock on the shelf, and if X's impressions were found to continue unchanged when the clock was removed or its position altered, then I think we should be inclined to say that X did not see the clock which was before his eyes, just because we should regard the clock as playing no part in the origination of his impression.
>
> Or, to leave the realm of fantasy, it might be that it looked to me as if there were a certain sort of pillar in a certain direction at a certain distance, and there might actually be such a pillar in that place; but if, unknown to me, there were a mirror interposed between myself and the pillar, which reflected a numerically different though similar pillar, it would certainly be incorrect to say that I saw the first pillar, and correct to say that I saw the second; and it is extremely tempting to explain this linguistic fact by saying that the first pillar was, and the second pillar was not, causally irrelevant to the way things looked to me.   [pp. 461–462]

The value of these two arguments, however, is open to question when we consider the possibility of intuiting abstract objects. Consider the latter argument, which depends on the possibility of numerically different but qualitatively similar objects. This possibility does not seem to apply to the case of abstract objects. Anything "similar" to a mathematical object would be identical to it. The distinction often drawn between qualitative and numerical identity appears to lapse in the case of mathematical objects and structures.

Mathematical Knowledge

The relevance of Grice's initial argument, furthermore, must also be proved when we apply it to the case of mathematical objects. In the case of physical objects, such as a clock, being in a certain perceptual state, even the state of "being appeared to" (Chisholm's phrase) as though there were a clock, is indeed not sufficient to be seeing the clock. The possibility of "true hallucination" or illusion is present: even if the perceptual state corresponds with the true state of affairs, the perceptual state is not that of "seeing," so long, in Grice's words, as that state of affairs is not "causally relevant" to the sensory impressions. But is it so obvious that there is a distinction, in the case of the intuition of abstract objects, between "true hallucination" and perception? Does not the distinction here collapse? As Descartes says (Meditation I), we may be certain of mathematical truths even while dreaming. Why is it not enough for perception (assuming it exists) of abstract objects to be in the "right" perceptual state, unlike seeing, where the "right" sensory impressions must be caused by the "right" event (thing) (suggested by Baruch A. Brody). This question is rhetorical, for in the following pages an attempt will be made to show that there is a difference, even in the case of mathematical intuition, between "true hallucination" and perception. But it is far from clear at this stage that the difference must be marked out with reference to condition (C). It may well be that (C) holds only for entities that make a causal difference to us.

(It seems worthwhile noting, by the way, that we have not touched an obvious version of the CTP, one, incidentally, hinted at by Grice in his talk of "causal relevance," according to which an object cannot be seen unless a *sentence* containing a name of it must appear in a causal explanation of my relevant sensory impressions. This version, of course, is perfectly compatible with mathematical intuition even of objects. This causal condition, however, "quantifies over" linguistic entities, and thus is heir to many problems: When do words "exist"? What is a sentence? Since I do not have answers, I forbear.)

In sum, whether the causal theory makes mathematical intuition impossible depends on a negative answer to several moot questions:

Can one (*perceptually*) *see that p* without, for any relevant x, seeing x? Can one see that $2 + 2 = 4$ without "seeing" 2?

Is there a plausible version of the causal theory of perception which, unlike (C), permits the "seeing" of abstract entities?

Is (C) true—or must an exception be made in the case of abstract entities?

I am far from able to treat these questions as they deserve to be treated. But perhaps I have sown some doubt that the answer to all the questions is "no." Platonism would be in much better shape, however, if the existence of mathematical intuition were itself a viable proposition. A priori objections have a habit of paling in the face of *faits accomplis*. And even if we could dispose finally of all objections, we would still need an account of mathematical intuition. Clearing a site is hardly building the house.

It is time, then, to turn our attention more concretely to mathematical knowledge and to mathematical intuition as a possible vehicle of such knowledge.

IV

We have said much in defending mathematical intuition; but little to characterize it. The critic of epistemological Platonism may thus demand either a characterization (preferably together with evidence that something is characterized thereby) or else proof that no characterization is necessary. We have said only that mathematical intuition purports to be a faculty on the analogy of sense perception for acquiring mathematical knowledge. We have also noted that a Platonist might go whole hog (like Plato) and claim for himself or for another direct apprehension of the mathematical objects themselves (on the model of "seeing an *F*"); or he might content himself with "intuitions that" certain propositions are true—stoutly maintaining, however, that these intuitions are perceptual, even if no *thing* is ever perceived. Finally, we quoted a passage from Gödel in which he expounds the position, but his words leave much unsaid.

First, we want to know how much rests on mathematical intuition, according to Gödel. For example, mathematical intuition is sometimes used to support *ontological* Platonism, as evidence for the existence of mathematical objects. Gödel seems to accept this point of view, for he introduces the concept of intuition to give content to undecidable propositions such as the continuum hypothesis. Sets exist, he says, because we intuit them (just as physical objects exist, because we see them) and the existence of such objects renders intelligible questions which are today undecidable. Thus Gödel considers the question of ontological Platonism "an exact replica of the question of the objective existence of the outer world" (1947, in *BP*, p. 272), presumably because if one doubts the existence of mathematical objects that one comes to know through intuition, one must similarly doubt the existence of material bodies that are known through the senses.

This, of course, makes mathematical intuition hold up a lot, perhaps more than it can bear. We do not even know yet whether, even granted that sets exist, we could apprehend them, yet Gödel is making such intuition the cornerstone of his view of mathematical truth. How, indeed, are we to distinguish mathematical "intuition" from hallucination; what makes it "veridical"? We seem to be headed toward mysticism.

There is a more modest way of regarding mathematical intuition. This is to argue independently for ontological Platonism, without any recourse to mathematical intuition initially. One would show (in the manner of Quine) that the existence of mathematical objects is guaranteed by their indispensability to science, by our inability to say what we want to say about the world without quantifying over them. Having convinced ourselves of the existence of mathematical objects, the next step would be to explore how one in fact comes to know mathematical truths.

Is mathematical intuition a live option in explaining mathematical knowledge? George Berry (1969, p. 261) thinks not:

> How then do we find out about this realm of extra-mental nonparticular, unobservable entities? Our knowledge of them, like our knowledge of the extra-mental, unobservable objects of the physical sciences, is indirect, being tied to perceived things by a fragile web of theory. In both cases—physics and logic—our hypotheses about the unperceived are tested by their success in accounting for the character of the perceived. Misreading this similarity, one might easily conclude that a faculty of non-sensory perception, call it 'intuition', is necessary to play a part in logic parallel to the role of sensation in physics. The conclusion is groundless. Long-run success in dealing with the same old perceptual field of ordinary sensation holistically confirms not only our belief in a force satisfying an inverse-square law but also, if more remotely, our belief in the derivatives used to compute the force. It also confirms our belief in the classes ultimately invoked to so analyze the derivatives as to explain the computations.

Now this point is all right as far as it goes; the argument for mathematical intuition is pretty weak if it rests only on an analogy. But Berry has not disposed of mathematical intuition merely by refuting a bad argument. The question remains: "How do we find out about this realm of extramental . . . entities?" According to Berry, the answer seems to be: By experiment. But Berry has asked: How do we find out, i.e., come to know? And it seems fantastic that the way one learns about the properties of classes is by predicting one's sensations on the basis of the sum-set axiom. I am not suggesting that Quine is wrong in thinking that the ultimate court of appeal for the sum-set axiom is the holistic considerations that Berry mentions. But such considerations obviously play no role in the mind of the average mathematician, who may never have seen the reduction of mathematics to set theory, but who pursues set theory for its own sake. Such a mathematician is able to "find out" all sorts of things about abstract entities by a method we as yet do not understand. That holistic considerations and semantic ascent are the ultimate ground of mathematical knowledge does not eliminate the role of intuition—on the contrary.

Consider someone who, asked how he knows that there are gravitational fields, answers that he feels them, or someone who claims to detect the presence of an electron by viewing its trail in a cloud chamber. These reasons are valid, but they are valid only given an entire theory that justifies thus describing "what one sees." Or take Quine: he spends the first pages of *Word and Object* justifying his assumption of the existence of "middle-sized objects" by the same sort of theoretical moves that he later uses to legitimize quantification over sets. Given the theoretical background, however, *seeing* (material bodies) is enough for believing (that they exist here, now). It is the theoretical considerations that, for Quine, justify our speaking as though we saw physical objects. The reader may be reminded here of Carnap's "Empiricism, Semantics, and Ontology" (1956), and the distinction between "internal" and "external" questions, though the distinction is here transferred to the realm of epistemology. And even Quine, who derides the Carnap distinction between "internal" and "external" questions in ontology, nevertheless speaks of the difference between the epistemological "job" of the mathematician, and that of the "ontologist" (i.e., Quine):

What distinguishes between the ontological philosopher's concern [and the concerns of the scientist] is only breadth of categories. Given physical objects in general, the natural scientist is the man to decide about wombats and unicorns. Given classes, or whatever broad realm of objects the mathematician needs, it is for the mathematician to say whether in particular there are any even prime numbers or any cubic numbers that are sums of pairs of cubic numbers. On the other hand it is scrutiny of this uncritical acceptance of the realm of physical objects itself, or of classes, etc., that devolves upon ontology.   [*WO*, p. 275]

Even this holistic division of labor still presupposes that we can distinguish between saying *that* classes are and saying *what* they are—what kinds there are. This is a position Quine has already rejected, saying:

> One tends to imagine that when someone propounds a theory concerning some sort of objects, our understanding of what he is saying will have two phases: first we must understand what the objects are, and second we must understand what the theory says about them. In the case of molecules two such phases are somewhat separable [but] in the case of the wavicles there is virtually no significant separation; our coming to understand what the objects are *is* for the most part just our mastery of what the theory says about them. We do not learn first what to talk about and then what to say about it.
>
> Picture two physicists discussing whether neutrinos have mass. Are they discussing the same objects? .... The fact that both physicists use the word 'neutrino' is not significant. To discern two phases here, the first an agreement as to what the objects are (viz. neutrinos) and the second a disagreement as to how they are (massless or massive), is absurd.   [*WO*, p. 16]

In the case of set theory, to uphold a distinction between the "ontologist" and the mathematician would more or less relegate the latter to deducing the consequences of various axioms chosen by the "ontologist." If the mathematician is allowed to choose the axioms too, how could his activity be distinguished from finding out whether to accept sets as "given"? Quine himself states in "Carnap and Logical Truth" that one cannot segregate out various preferred axioms as somehow giving the essence of classes; in adding new axioms, or changing old ones, we cannot say whether we are discovering new truths about a previously "given" category or whether we are scrutinizing the "category" itself (see *WP*, pp. 106–107).

Mathematical Knowledge

It seems to me, therefore, that Quine's epistemological version of the "internal-external" distinction is subject to the same criticism as that which he himself directs at Carnap. The distinction I wish to draw, however, does not rest at all on the different questions one might ask about a subject matter. It is a distinction between various questions one might ask about someone's *knowledge* of the subject matter.

For example, depending on the level of the question "How do we find out about physical objects?" there are two answers. One might reply, "By looking at them," or else by the holistic arguments that open *Word and Object*, whose import is that assuming the existence of physical objects helps in (is indispensable for) predicting future experiences. The answers to the question are connected in that the strength of the former depends upon that of the latter. Even statements concerning "middle-sized objects" are not reports of sense data; they too, for Quine, are "theoretical." So only after accepting the utility of "quantifying over" material bodies can we, in retrospect, accept observational reports of such bodies. Quine would certainly not wish to say that knowledge concerning bodies is limited to those who know the holistic calculation. But the possibility of such calculation is what justifies perceptual knowledge concerning those bodies, in individual cases. Knowledge may be based on reasons not accessible to the knower.

The case could be the same for mathematical intuition. Intuition there might be, though guaranteed by holistic considerations. Such an intuition, however, would provide a degree of certainty for mathematical knowledge no greater than that provided by the holistic arguments. But surely, classical Platonism attempted to give an account of the seeming certainty, the indubitability, of mathematical truths, by invoking intuition. Can the holist, with his weakened intuition, account for this certainty?

The ontological Platonist finds this an acute problem. For him, mathematics is the science of abstract entities, just as atomic

Platonism and Mathematical Knowledge

physics is the science of micro-entities. Abstract entities, if they exist, are more remote from direct sense perception than even micro-entities. The corpus of mathematical theory would thus seem to be more easily revisable than any other, just as the latest theories about positrons are more easily revisable than the laws concerning metals at high temperatures in a vacuum. By the same token, mathematics would be less likely than any other theory to come into conflict with observation. But this hardly makes for *certainty* in mathematics, any more than in positron theory.

Quine is aware of the problem posed by mathematics to his holistic epistemology and addresses it directly in two places. In the introduction to his *Methods of Logic* (1959), having first sung the authority of observational truths, he adds:

> There is also, however, another and somewhat opposite priority: the more fundamental a law is to our conceptual scheme, the less likely we are to choose it for revision. When some revision of our system of statements is called for, we prefer, other things being equal, a revision which disturbs the system least. . . .
>
> Our system of statements has such a thick cushion of indeterminacy, in relation to experience, that vast domains of law can easily be held immune to revision on principle. We can always turn to other quarters of the system when revisions are called for by unexpected experiences. Mathematics and logic, central as they are to the conceptual scheme, tend to be accorded such immunity, in view of our conservative preference for revisions which disturb the system least; and herein, perhaps, lies the "necessity" which the laws of mathematics . . . are felt to enjoy.   [p. xiii]

In *Philosophy of Logic*, Quine spells out this "centrality."

> Mathematics has its favored lexicon, unlike logic, and its distinctly relevant values of variables. But, despite all this, mathematics presents as impartial a front to natural science as logic does. For the distinctive terms and the distinctive objects of mathematics tend mostly to favor one branch of natural science no more than another. . . .
>
> . . . Finally . . . mathematics is . . . a handmaiden to some extent of all natural sciences, and to a serious extent of many. We might say at the risk

of marring the figure that it is their promiscuity, in this regard, that goes far to distinguish logic and mathematics from other sciences.

Because of these . . . traits of logic and mathematics—their relevance to all science and their partiality toward none—it is customary to draw an emphatic boundary separating them from the natural sciences. [p. 98]

Not that Quine is in agreement with the emphatic boundary. Nor does he think that the truths of mathematics are directly supported by observation:

When I speak for a kinship between mathematics and natural science I do not mean this. The ' + ' of '7 + 5' should connote no spatial assembling of objects. . . .

The kinship I speak for is rather a kinship with the most general and systematic aspects of natural science, farthest from observation. Mathematics and logic are supported by observation only in the indirect way that those aspects of natural science are supported by observation; namely, as participating in an organized whole which, way up at its empirical edges, squares with observation. [*PL*, p. 100]

Mathematics and logic are *obvious* only because "relevant" to all science and "partial" to none, not because of their direct observational support. Quine would charge Mill with being misled by the appropriateness of mathematics to all sciences and to everyday life.

But how is it that mathematics—the "natural history" of abstract objects (the Wittgensteinian figure)—and logic are so ubiquitous? As far as logic, Quine's own explanation is that logical truths have this salient trait: "All other sentences with the same grammatical structure are true too" (*PL*, p. 101). Logical truths, to put it another way, are "invariant under lexical substitutions" (*PL*, p. 102), thus applicable to any context, "topic neutral."

Quine neglects to explain, however, the ultra-relevance of mathematics, which cannot be accounted for as he did logic. Mathematics has a vocabulary of its own, as Quine says: the addition sign, the integral sign, and the like. Mathematical truths are not invariant under lexical substitutions; they also presuppose a domain of abstract objects, unlike logic.

An account of mathematics, however, is not difficult. To apply mathematics, we need only broaden its vocabulary to include terms from the other sciences. Set theory can be applied by adding new constants denoting various individuals studied in chemistry or anything else; the theory of functions can be extended to encompass functions from domains of material bodies merely by adding vocabulary. Such extensions, for example, enable us to speak of the *number of* a set of material things and apply the results of pure arithmetic to these numbers—the same numbers as are studied in pure arithmetic. The dichotomy "pure" vs. "applied" arithmetic is false if intended to mark a distinction in the subject matter. Only a pseudoproblem, therefore, lurks in the "applicability" of mathematics to the world. '3 + 2 = 5' means precisely the same in pure mathematics as it does in applied.

Extending function theory to include functions from physical domains enables us to apply analysis to nature. Physical magnitudes can often be seen as functions from material bodies into the real line, or into space (n-space being the Cartesian product, n times, of the real line with itself)—functions, that is, relative to various parameters, such as a chosen coordinate system of bodies or a standard unit of length (e.g., a body stipulated to be a meter long). The equations that link such magnitudes, for given parameters, are relations linking the values of these functions—thus relations on the reals, or on n-space. Where these relations are, in fact, expressible in mathematical vocabulary (as, for example, in equations) the machinery of analysis yields physical insight. This happens when we "solve" a differential equation. We cannot know a priori, to be sure, that the various physical magnitudes will be thus simply related. Some scientists have even expressed astonishment that this is so as often as it is (Wigner); Quine has attempted to dissolve this feeling (1966a). However this may be, the vocabulary in which these relations are expressed is precisely

129

that of pure analysis. Ultimately, then, the universal applicability of mathematics results from the fact that items of any kind may be gathered into a domain, a totality, a class.

The holist argues correctly that a single change in mathematical theory would result in wholesale chaos in scientific laws across the board. Applying the "maxim of minimum mutilation" to this situation explains why mathematical truth seems so certain.

One objection to holism, that it cannot account for the indubitability of mathematics, is thus inconclusive. The traditional doctrine of intuition attempted to bridge the gap between the human and the Ideal by postulating a special human faculty lodged in the soul. It attempted to reconcile the remoteness of mathematical objects with the firmness of mathematical knowledge. Thus also Gödel (1947, in *BP*, p. 272). But intuition turns out not to be necessary for the reconciliation.

With all this, the holist cannot yet explain how human beings *come* to know mathematical truths, though he can justify such knowledge ex post facto. Perhaps intuition may be of use here.

V

Assuming the truth of ontological Platonism, we thus proceed to explore the usefulness of mathematical intuition to mathematical epistemology. Our remarks will be speculative; few positive suggestions will be made. But something has to be said— as we shall see, many philosophers seem to take mathematical intuition seriously.

Claims to intuition appear in two forms, as we have seen. There is the alleged intuition of objects, and there is the alleged intuition of truths. Semantically, there are two expressions to be studied: "intuition of x," 'x' a variable; and "intuition that p," 'p' a schematic letter. (Philosophers who, like Church, believe in intensional objects, may let 'p' vary over propositions. For them, the intuition of truths, intuition-that-p, will also be a relation between the knower and an object, an intensional object.)

Platonism and Mathematical Knowledge

In our tough-minded age, it is strange to hear philosophers speak of "intuition." In the philosophy of mathematics, however, such talk is frequent, and heard in the best circles. Listen to Parsons: "The same kind of intuitive construction as is involved in arithmetic is also involved in *perceiving the truth* of logical truths" (1965, p. 203; see also 1971). Tharp, in a perceptive article, lets slip the following aside: "We are treading on very slippery ground indeed. I of course cannot pretend to give an account of the 'perception' of mathematical truths. Yet one feels that the heart of the matter lies here, and, for the question at issue, one need not provide such an account" (1971, p. 162). And there are others (see Chihara, 1966, sec. I, for some names; see also Pollock). Like Tharp, few of these philosophers ever supply an "account" of mathematical intuition. (Parsons, 1971, "Mathematics and Ontology" attempts an account, on Kantian lines.)

No one today, however, upholds hard-core intuition—the direct intuition of mathematical *objects*, the first type of mathematical intuition mentioned above. No one, with the possible exception of Gödel on one reading of the passage quoted above, claims to having perceptions of individual numbers. The reason seems to be that there is no information to be gained from "looking" at an individual number. Benacerraf (1965, cited above in Chapter II) may be wrong in his conclusions, but he is certainly correct that the numbers have no properties but those they have in relation to one another.

But how can 'intuition-that', the remaining and defensible kind, present any interest to a philosopher? Either we have the ability to conjecture correctly, or not. How can mere true conjecture be knowledge? First, conjecture is fallible; even the greatest masters of mathematics failed at it. So the blanket, oracular authority of a conjecturer cannot be invoked in support of the knowledge-claim, even by a third party (for we have already observed, that a person can *know* on the basis of reasons he is not, but others are, in position to give). If present at all, intuition is present in only some cases of true belief, where there is a special sort of explanation that accounts for the conjecture in question. We need to be able to "explain away" failures in order that the true intuitions will be uncontaminated by the surrounding ignorance, just as we explain away hallucinations in order not to cast doubt wholesale on our remaining perceptual beliefs.

To speak of intuition, therefore, is to offer a promissory note: the claim that an explanation of a certain kind is available for certain of the true mathematical beliefs we and mathematicians have, an explanation that will explain also why certain of our other mathematical beliefs can be expected not to be vindicated. It is to postulate a certain mental faculty. (I stress again that by "explanation" I do not mean one necessarily accessible to the knower.)

Not any explanation will do in giving an account of intuition, because it may well be that all our beliefs have explanations, even lucky guesses. Nor is it enough to say that the explanation in question must utilize the fact known, for we have seen that every explanation of anything is likely to use the axioms of arithmetic or of analysis. The explanation must do more than explain the origin of a belief. It must also show that the belief is trustworthy, and how.

Whether a perceptual belief is trustworthy depends not only on the circumstances of the individual belief, but on the reliability of the method used in reaching the belief. Consider the example of a machine, wired fortuitously to produce an hallucination of itself, precisely where it is in fact standing. Someone attached to the machine, unknowingly say, would not count as knowing where it is, if his only basis were the experiences he was being stimulated to have. And this, even though the machine itself was responsible for the induced experience. For had the machine been moved slightly, the lucky blending of truth with hallucination would have been cut off. A machine, on the other hand, that produced experiences generally corresponding to reality would be considered a mechanical eye.

Intuition, of course, is not considered to be initiated from without but self-induced. Still, intuition must pass tests as did the machine. It would have to be a repeatable technique. There would further have to be a method of checking up on intuitional knowledge to establish whether the technique had been applied correctly or not. (I am of course indebted to Wittgenstein's *Philosophical Investigations* for the general approach.) Otherwise, there is again no distinguishing fortuitous conjecture from believable intuition in those cases where the conjecture comes up true.

The difference between mathematical intuition and physical perception might *seem*, as we saw above in the first section, to be this. A person might undergo the *same* sensations: once as perception, once as hallucination. Surely the same sensations may be caused either by normal seeing or as a result of monkey business. Intuition *seems* different as we have seen in the preceding section: the concept of "hallucination" seems to lapse; even in dreams one might come to know mathematical truths, tricky demons aside. It might be considered that no "checking up" is necessary: either a subject is in a self-verifying state of intuition or not.

I would argue (still in the manner of Wittgenstein) that even a proper mental state is not enough for knowledge. The knower must have the power to discriminate between the true state, and bordering false ones. If a person does not have this ability to check up on himself, he is in no position to know. For were he confronted with intuitions of something else, he might not know the difference.

Here is where the "checking procedure" come in. If the knower has a method of ascertaining whether the intuitions he has are really, say, those of Hilbert space, or whether the process, the mental process, that has culminated in the present intuitional state is a reliable one, we might credit him with mathematical credentials of knowledge.

VI

We have spoken of "checking procedures." But if we are speaking of propositional intuition, intuition that p, do we need a separate procedure for each individual truth?

To avoid this absurdity, we might view with Benacerraf (1965) the point of mathematics as the study of "structures," rather than individual mathematical objects. The point, however, is epistemological rather than ontological—we accept mathematical objects, contra Benacerraf, but we agree that the only things to know about these objects of any value are their relationships with other things. (This is the mark of abstract objects.) Intuition becomes then the intuition of structures rather than the intuition *either* of truths or of individual objects. This new point of view has the virtue of conforming to the way mathematicians speak of intuition. One speaks of set-theoretic intuition, analytic intuition, geometric intuition, topological intuition, and so forth. Intuition in one branch of mathematics, furthermore, is alleged not to go with intuition in another.

W. Sellars has suggested that sensations be regarded as theoretical entities modeled on physical objects. To have a sensation of a red triangle, according to him, is to be in a psychic state that bears an analogy (in the sense of a scientific model) to a red triangle (1963, pp. 190–194). It *may* be illuminating to think of mathematical intuition as a state of the perceptive faculty that bears a partial analogy to a mathematical structure. This is done perhaps through what Locke called "abstraction." One imagines or looks at material bodies, and then diverts one's attention from their concrete spatial arrangement. One gathers up in one's mind the objects into a manifold, and then has an intuition of their structure. This is how one might become familiar with the standard model of ZF set theory—by abstracting from dots on a blackboard arranged in a certain way. Thus one arrives at an intuition of the structure of *ZF* sets. (See Boolos [1971] for a rigorous development of *ZF* set theory from the intuitions thus obtained.)

The result of the abstraction process is not to be regarded as an intuition of the individual sets that make up the *ZF* ontology. It is an intuition of their structure. Having an intuition of this structure, moreover, is not the same as imagining the array of blackboard dots "in no particular order." (Berkeley would have nasty things to say about such an idea.) Having an intuition of the structure of *ZF* sets may begin with imagining dots in an array, but it does not end that way. Once one has abstracted from the geometric features of the array, the mental state is qualitatively different from simple imagining. It is modeled on the structure of abstract objects, not physical objects or a physical array. When we speak of "diverting one's attention" or "abstracting" from specific features of an image, these expressions should be regarded more as instructions for the perceiver, rather than descriptions of the end result, as in Locke (a weakened empiricism can accommodate abstract ideas, saving itself from the incoherence of Locke's account).

Although the above is only a feeble attempt to characterize mathematical intuition, I believe it points a way toward an empirical study of this faculty. I have tried to show only how a human faculty of acquiring mathematical knowledge might exist without running afoul of some standard objections. I conclude with some facts that point to the actual existence of such a faculty.

First, there is the extraordinary *agreement* among mathematicians of all cultures on the basic truths of mathematics. These truths can be checked independently, using holistic arguments, as above. But since most mathematicians do not have access to such arguments and many do not care about the efficacy of mathematics in science, one must conclude that the presence of a mathematical-intuitive faculty has some plausibility. Whatever explains this agreement will very likely also show that the agreement is based upon knowledge. It may be that this agreement is brought about by the relationship of mathematical truth to the evidence of physical sensation; but the way I see mathematical intuition, as beginning with sensory data, that relationship need not dishearten the Platonist. On the other hand, it seems totally false that the agreement among mathematicians of such diverse cultures should be attributable to universal training, as Wittgenstein thought, though I leave the proof of this to another occasion.

Second, there is evidence that the great mathematicians have been able to convince themselves of the truth of various mathematical propositions without knowing their proof. The Indian mathematician A. K. Ramanujan is perhaps the most striking example. Completely unlettered in Western rigor, even later exposure to Western standards of proof did him no good, he had the ability to conjecture the most complicated formulas in the theory of elliptic functions, many of which were later proved. The "mathematical," esthetic, quality of his conjectures showed that more than mere guesswork was operating, as did his improbably high percentage of success. As the English mathematician G. H. Hardy, Ramanujan's Western discoverer, put it on examining the latter's notebooks filled with the most extraordinary assertions, Ramanujan must be a genius, since these are more frequent than charlatans of such skill (see Newman, 1956). Perhaps mathematical intuition was at work in Ramanujan, but investigation is necessary. We have already seen, in the case of Euler, that inductive methods can lead to knowledge in mathematics. Ramanujan, a prodigious calculator, had access to mathematical data that the normal mathematical scientist could not have, at least in the precomputer age. For what it is worth, however, my own belief is that Ramanujan did have a perceptual feel for mathematical structures.

Finally, we should not ignore the utterances of mathematicians themselves. We may beware of introspection, but it would be foolish to ignore it. Mathematicians do claim intuition for themselves: they say that intuition grows with familiarity with the subject matter (e.g., Cohen, 1966, pp. 150–151) much as our sense perception improves through familiarity with a terrain. We lack at least an explanation of the conviction of so many of the best mathematicians.

What is striking about modern mathematics is that its proof procedures often give little insight into the subject. Formalized arguments are deliberately constructed to amaze the public with tours de force and to conceal the intuitive grounds upon which the theorem originally rested. Proofs in analysis often show that certain facts are true, without showing *why* they are true. (Compare a standard work in analysis with the arguments in Courant and Robbins, 1941, for example.) The same holds for topology, where absence of intuition will make even some of the proofs incomprehensible because unsurveyable in a subtle way.

My excuse for these remarks is a feeling of responsibility to a topic that many have skirted but few have faced. I have attempted to focus light on a concept usually deliberately obfuscated, both by its proponents and by its detractors. I have tried to dissolve some of the main confusions regarding mathematical intuition, so that its empirical study can become possible.

Appendix to Chapter One

Mathematical Induction and Mathematical Knowledge:
Poincaré's Critique of Formalism

Chapter One toured the battlefields of logicism, the thesis that mathematics might be based epistemologically upon logic. The classical charge against logicism of circularity was examined and rejected, though other objections were sustained. In this addendum, I should like to assess the worth of similar charges of circularity made by Poincaré against another great mathematical epistemology, Hilbert's formalism.

"One cannot know a mathematical truth unless one knows a proof of it." Though this statement is false (see Chapter Three), Hilbert would certainly have agreed, with exceptions. For Hilbert conceded to the intuitionists that there is a core of mathematical truth that needs no verification other than our intuitive grasp of its truth. This core of propositions deals with "certain extralogical concrete objects that are intuitively [*anschaulich*] present as immediate experience prior to all thought" ("On the Infinite,"

1926, p. 376). Among these propositions are all numerical equalities of the form

$$\mathfrak{a} + \mathfrak{b} = \mathfrak{b} + \mathfrak{a}$$

where the constants designate numerals in the stroke system—/, //, ///, and so forth. These propositions he calls "contentual" (*inhaltlich*). The universal proposition

$$(y)(x)[x + y = y + x],$$

on the other hand, is not part of the contentual core (p. 380).

Hilbert's formal system, however, contains the means for generating truths which are noncontentual, though some of the theorems correspond to contentual truths. For these "nonfinitary" truths (and Hilbert identifies '*inhaltlich*' with 'finitary'), it is necessary to prove them formally from the axioms of the system. Hilbert would have agreed, then, with our dictum, provided we limited it to noncontentual truths. (Actually, Hilbert might have objected to our calling them "truths.")

Indeed, Hilbert accepts something stronger. He places great emphasis on a *consistency* proof for mathematics, the axioms and proof procedures. Only given a finitary consistency proof can we be assured that we can trust the formal proofs themselves. To know a noncontentual mathematical truth, therefore, Hilbert requires knowing *two* proofs: a proof of consistency and the proof of the theorem.

But here we must confront Poincaré's objections, or objections in his spirit. (Poincaré in fact criticized an earlier system of Hilbert, which did not include mathematical induction. See Hilbert, "On the Foundations of Logic and Arithmetic," 1905.) What good, after all, is a consistency *proof*? Since the proof unavoidably will use mathematical induction (itself an axiom of the formal system), the proof cannot make us any more certain of the consistency of the system than the certainty we place in mathematical induction. But if we are certain that mathematical induction will never lead us astray, why a consistency proof in the first place? Of course, one might reply: There are other axioms in the system besides the axiom of induction. Hilbert, in fact, considered the "logical epsilon axiom,"

$$A(a) \rightarrow A(\varepsilon(A)),$$

as the most problematic (1926, p. 382); from this axiom can be derived all the principles that make intuitionists turn blue, for example, that if a predicate is not true of everything then there must be a counterexample:

$$-(a)A(a) \rightarrow (\exists a) - A(a).$$

Nevertheless, Hilbert did not avail himself of this reply. He responded thus:

> Poincaré already made various statements that conflict with my views; above all, he denied from the outset the possibility of a consistency proof for the arithmetic axioms, maintaining that the consistency of the method of mathematical induction could never be proved except through the inductive method itself. But as my theory shows, two distinct methods that proceed recursively come into play when the foundations of arithmetic are established, namely, on the one hand, the intuitive construction of the integer as numeral . . . that is, *contentual* induction, and, on the other hand, *formal* induction proper, which is based on the induction axiom and through which alone the mathematical variable can begin to play its role in the formal system. . . . Poincaré often exerted a one-sided influence on the younger generation. ["The Foundations of Mathematics," 1928, pp. 472–473]

What is the distinction here? It appears to be the distinction between inductive *procedures*—i.e., building up a chain of proofs to conclude first 'A[0]', 'A[1]', 'A[2]', etc., as long as we please, and the use of the induction *axiom* to detach universally generalized propositions. The consistency proof, it is claimed, uses only inductive procedures, but not an axiom of induction. The proof shows, of any given formal proof, that it is not the proof of 'p & −p'. For we can perform the replacement of numerals for quantifiers outlined for us by the consistency proof as far as we want to go.

But this distinction itself needs clarifying. Surely the consistency proof, to give us what we need, must assure us of *every* proof that it is not a proof of 'p & $-p$'. This fact is obscured by phrases like "any particular proof." We still need an axiom of induction in the metalanguage, to assure ourselves indeed that the construction will always terminate, however long the proof. True, for proofs of length 7 or length 1,000,000 or any particular length we do not need an induction axiom if we are willing to carry out the numeral replacement. But the consistency proof speaks of proofs of "any" length, and what is the difference between that and proofs of "every" particular length?

To guarantee that the inductive *procedure* works for *any* given integer, one needs an induction *axiom* of a higher order. So if Hilbert eschews an induction axiom in his metalanguage consistency proof, he will need a meta-metalanguage induction axiom to guarantee that the metalanguage procedure works in general.

Of course, there is a valid distinction between object language principles and principles expressed in a metalanguage, but it is not relevant here. What Poincaré and we want to know is why a certain principle is epistemically more suspect in one language than in another.

It is possible that the distinction Hilbert has in mind is between types of properties that can be studied using mathematical induction (Thomas Scanlon, Personal Communication). The induction axiom has the form

$$(F0 \mathbin{\&} (x)(Fx \supset FSx)) \supset (x)Fx$$

and any property, however noncontentual, is a proper value of 'F'. What Hilbert is saying, according to this, is that the consistency proof makes use of only constructive instances of the induction principle.

But it seems to me that the passage quoted above does not

suggest this distinction. First, this interpretation of Hilbert seems to leave unexplained Hilbert's remark about the "role" of the "mathematical variable" in the passage. Second, if this is what Hilbert means, his point is not about *induction* per se. For whatever the property denoted by '*F*' in the induction axiom, the induction principle is as *inhaltlich* as ever. The principle states only that *if* a certain property is true of 0 and *if* it holds of the successor of any number of which it holds, then it is true of any number whatsoever. If the property is nonconstructive, we may not be able satisfactorily to verify that it is indeed true of 0 or that it is hereditary, and the induction will never get going. But this does not make the induction nonconstructive, just the property. Given a nonconstructive property, the instance of the induction axiom as a whole might strike us as meaningless; but this is not peculiar to induction, it is also true of the other axioms when they are forced to include nonconstructive properties. So while there are certainly metamathematical differences between a system that allows "nonconstructive induction" and one that allows induction only on "constructive" properties, this fact does not seem to imply that there are two types of inductional procedures. It is rather that in so-called "nonconstructive" induction the induction bases are more shaky because nonfinitary.

I feel that the truth lies in the original insight, as yet undeveloped. This was that the consistency proof shows of any number that it is not the length of a proof of a contradiction, while the use of induction in general is designed to show of *every* number that a certain property holds. But I shall no longer suggest that the distinction is between the use of an inductive procedure and the use of an induction axiom; rather it is a distinction between two uses of an induction axiom. In one use, I shall argue, the mathematician shows that a property holds for every value in a completed totality of values. In another use, he shows that a property holds for *any* value, in a totality that need no longer be considered completed.

Appendix to Chapter One

If there is such a distinction, it will give also a key to understanding another Hilbertian dichotomy, that between

$$\mathfrak{a} + 1 = 1 + \mathfrak{a}$$

and the corresponding algebraic proxy

$$a + 1 = 1 + a.$$

The German letter equation is interpreted by Hilbert not as a "combination, formed by means of 'and,' of infinitely many numerical equations," but only as "a hypothetical judgment that comes to assert something when a numeral is given" (1926, p. 378). The algebraic formula, on the other hand, with its variables, is not contentual; its negation is well formed and purports to say that every single number commutes with 1.

That this distinction is the key to understanding Hilbert's two kinds of induction may be seen by his remark, quoted above, that formal induction is the means through which alone "the mathematical variable can begin to play its role in the formal system." Hilbert means here that without the use of a variable we can proceed from $F0$ to $F1$ to $F2$, etc., and (from the contentual point of view) might even be able to enunciate the axiom

$$(F0 \mathbin{\&} (x)(Fx \supset FSx)) \supset F(\mathfrak{a})$$

but not the full result

$$(F0 \mathbin{\&} (x)(Fx \supset FSx)) \supset Fx.$$

In other words, induction allows us to insert a variable where before we could only insert numerals or at most German letters. In this respect, it is a more powerful principle than universal generalization.

Still the distinction is not yet clear. A "hypothetical judgment" is a queer fish. If we try to say instead that the German letter equation states, in effect, that every substitution instance is a truth (German letters are schematic numerals rather than

Mathematical Induction

variables), we will still in the metalanguage have to use a true quantifier ranging over an infinity of *numerals* when we try to say what we know, when we know a German letter equation. But let us press on.

We said that contentual induction yields only that for any natural number a certain property holds, while formal induction purports to show that the property holds for every number "at once." In other words, that the generalized result of "contentual induction" is a statement that does not presuppose a completed totality of numbers, while the result of "formal induction" is a generalized statement that does.

One might try to give the difference as Hilbert did between German-letter and variable-letter statements, by defining 'every' statements to have negations, with 'any' statements not to have them (1926, p. 378). But this criterion is not of much use, since it is not clear how to distinguish those universal statements that have and those that have no negation. We do not assert the negations of universal theorems proved by mathematical induction, for example; so how can we tell whether a theorem so proved is to be construed as an 'any' or an 'every' generalization?

Let us remember the Wittgenstein doctrine that the key to understanding the meaning of an expression is the expression's use, while forgetting his unfortunate dogma that the meaning of a mathematical statement is set forth in its proof only. Now one of the cardinal uses of a mathematical statement is as a premise in deductive inferences. If we can segregate inferences into those that demand the 'every' interpretation of generality and those that do not, we shall have a case for ambiguity in the universal quantifier. (Nothing turns on the word 'ambiguity'; a bifurcation of *use* would be sufficient for our purposes, whether or not the bifurcation creates an ambiguity in the used expression.)

I claim, then, that some inferences use generalized statements merely to generate specific instances, while other inferences make use of all the values of the bound variable. The former class of arguments, I say, demand only the 'any' interpretation; the latter, the 'every' interpretation.

If we look at the various things one can do with a universally quantified statement in a natural deduction system, we might at first conclude that the use of universal instantiation (UI) demands only the weak interpretation of the universal quantifier (since here, certainly, the universal is used only to generate instances) while all other inference types demand a stronger interpretation. Whenever the universal is used as such, without instantiating, the stronger interpretation ('every') is in question.

The facts are more complex than this. The use of a universal statement as such may be the result of a poverty-stricken vocabulary. To make this clear, I present a few examples of the use of generality in genuine mathematics. Consider first the argument from

$$(p)(0 \leqslant m/n \leqslant 1/p),$$

'm' and 'n' designating integers, to the conclusion

$$m/n = 0.$$

The argument proceeds in "natural deduction" as follows (many steps left out):

(1) $(p)mp \leqslant n$ from the premise

(2) $m = 0 \lor m \neq 0$

*(3) $m \neq 0$

*(4) $m \geqslant 1$ from (3) by arithmetic

*(5) $(n + 1)m \geqslant n + 1$ from (4) by arithmetic

*(6) $m(n + 1) \leqslant n$ from (1) by UI

*(7) $n + 1 \leqslant n$ from (5), (6), transitivity of $\leqslant$, commutativity of $\times$.

*(8) $1 \leqslant 0$ from (7) by cancellation

*(9) Contradiction   from (8) since we have $0 < 1$ and trichotomy

(10) $m \neq 0 \supset$ Contradiction   Conditionalizing

Mathematical Induction

(11) $m = 0$ (2) (10) by sentential calculus

(12) $m/n = 0$, qed   from (11) by arithmetic, dividing both sides by n.

The universal premise is here used only in step (6) to provide the counterexample, $n + 1$. Clearly, the weaker interpretation of the quantifier is all that is needed here. Had we proved the premise by mathematical induction, we would have consequently needed only "weak" mathematical induction: mathematical induction whose consequent—'$(x)\mathbf{A}[x]$'—is interpreted as "weak" generalization, used only to provide instances. Furthermore, if we knew how large n is, we could dispense entirely with the induction and prove directly that

$$(p)_{p \leqslant n+1}(0 \leqslant m/n \leqslant 1/p),$$

which would still yield the conclusion, $m/n = 0$. Of course, if we think of 'm' and 'n' as variables, the induction cannot be avoided. Even in that case, the point is that the induction is necessary only to produce "an arbitrarily large initial segment" of natural numbers so that, whatever n may be, we may instantiate for p one number—$n + 1$—to achieve the counterexample.

Consider, next, the analogous argument for the case of real numbers; the argument from the premise

$$(p)(0 \leqslant \alpha \leqslant 1/p), \alpha \text{ real}$$

to the conclusion

$$\alpha = 0.$$

The proof is similar, but the counterexample can no longer be generated constructively as before:

(1) $\alpha = 0 \lor \alpha \neq 0$

  *(2) $\alpha \neq 0$

  *(3) $1/\alpha > 0$ By the premise, α is nonnegative; by (2), the reciprocal exists

Appendix to Chapter One

$*(4)\ (\exists p)(p > 1/\alpha)$ Archimedes' theorem: for every real there is a greater integer

$*(5)\ (p) - (p > 1/\alpha)$ By premise and trichotomy

$*(6)\ -(p) - (p > 1/\alpha)$ (4), by conversion of quantifiers

$*(7)$ Contradiction   (5) (6)

$(8)\ \alpha \neq 0 \supset$ Contradiction   Conditionalizing

$(9)\ \alpha = 0$, qed.

In this deduction we do not instantiate the premise at (6). Our impulse might be, therefore, to invoke the stronger interpretation of the universal quantifier. The impulse would be premature: the reason we do not instantiate here is that Archimedes' theorem gives us no way of generating constructively the integer that is supposed to exist by (4). Nevertheless, we can think of the theorem as generating a counterexample nonconstructively: whether or not 'EI' is a basic or derived rule of our system. (In Quine's system, 1959, conversion of quantifiers is derived, and EI basic; and to go from (4) to (6) we should need to strip away the existential quantifier of (4) by EI, yielding the line $'-(p' > 1/\alpha)'$; but then we might as well forget about (6) and develop a contradiction by instantiating p' in (5).) If we had a richer object language with more function symbols, we could use the letter 'A', for example, to represent the Archimedean function, which picks out, for each real, the next higher integer (in mathematics, the symbol '[ ]' is often used, so that $A(x)$ would be $[x] + 1$, where x is not an integer). We conclude that the absence of UI is no reason to conclude that the universal quantifier bears the strong interpretation. (Also UI may be needed to get (5).)

It is the use of a universal statement as the minor premise in a *modus ponens* argument that requires us to assume the strong interpretation, or in analogous truth-functional inferences. In such cases, we make use "of all the values of the variable at once"; the completed totality of such values is what permits the argument to proceed. Consider the argument, so important to

Mathematical Induction

analysis (analysis is arithmetic for Hilbert, see Van Heijenoort, 1967, p. 383, n.6), from the premise

$$(n)(0 \leqslant a(n-1) \leqslant a(n) \leqslant b), a \text{ a real-valued sequence}$$

to the conclusion

$$(\exists b')(\lim_{n \to \infty} a(n) = b').$$

The proof is immediate from the least upper-bound principle

$$(1) \qquad (x)(Px \supset x \leqslant b) \supset (\exists b')[(x)(Px \supset x \leqslant b')$$
$$\& (b'')((x)(Px \supset x \leqslant b'') \supset (b' \leqslant b''))].$$

Letting 'Px' be '$(\exists n)[x = a(n)]$', and noting that the premise implies that

$$(2) \qquad (x)[Px \supset x \leqslant b]$$

we have, by *modus ponens*

$$(3) \qquad (\exists b')[(x)(Px \supset x \leqslant b')$$
$$\& (b'')((x)(Px \supset x \leqslant b'') \supset b' \leqslant b'')]$$

and the rest of the proof would prove that this least upper-bound, b', must also be the limit of the sequence.

In this proof, we really do use the premise in such a way that the weak interpretation is impossible. The consequent in (1) is contingent upon every *member* of P being bounded by b. To detach the conclusion at step (3), we thus need not merely an "arbitrarily long initial segment of values," but rather the totality of values. Conceptually, the least upper-bound theorem defines a function, operating on an entire set (possibly infinite) to produce one number—the least upper-bound of the set. Our first-order rendition of this axiom, as a schema, does not change these facts. In our previous example, the "suppressed" function was one whose argument was a single real number.

Our remarks about *modus ponens* jibe with the observation that, while 'Fx' may be substituted for '$(x)Fx$' in positive subformulas (barring captured variables, of course), this is not the

149

Appendix to Chapter One

case in negative subformulas, for example the antecedent of a material conditional. That is, the only way to express generality in the antecedent of a material conditional is to use, not suppress, the universal quantifier. (For $(\exists x)Fx \supset T$ says that T is true if any one x is F.) But in order to detach T from $(x)Fx \supset T$, we need to assert a minor premise, $(x)Fx$ with the same *significance* as the antecedent of the major, $(x)Fx$, hence one with a strong interpretation, unexchangeable for Fx. Our remarks, then, are also congruent with Hilbert's distinction between universal statements that do, and those that do not, have negations; for $A \supset B$ may be regarded as $-A \vee B$. It is true, as we asserted before, that we do not assert the negation of a universal theorem. But this asserted theorem must mean the same as the antecedent of the major premise, if the *modus ponens* argument is not to be a quibble; and the antecedent of this major premise, in turn, may be regarded as negated.

We conclude that only the use of a universal statement as the minor premise in a *modus ponens* argument requires us to interpret the universal quantifier as 'every' (or the use of the statement in parallel truth-functional arguments), where this universal quantifier is initial (we study initial quantifiers, of course, because only these are generated by mathematical induction). It little matters whether we impute ambiguity to the universal quantifier or only two different uses. For expository purposes, I continue to speak of two *interpretations* of the universal quantifier: the "strong" and the "weak" interpretations, where the former is imputed to universals allowed to figure as minor premises in *modus ponens* and equivalent inferences. We might write $(x)_s Fx$ as against $(x)_w Fx$, thus in effect having created two quantifiers. And finally, according as we interpreted the universal quantifier in

$$(F0 \ \& \ (x)(Fx \supset FSx)) \supset (x)Fx$$

as strong or weak, we get what I call strong or weak induction. Returning to Hilbert, my reading of his distinction between "contentual" vs. "formal" induction is that it is of a piece with the distinction between "weak" and "strong" induction. (Of course, Hilbert's system would have to be transformed into a system of natural deduction in order to vivify the two kinds of quantifiers, which are obscured in axiomatical logical systems—for *modus ponens*, in these systems, is used everywhere.) Weak induction is totally noncontroversial, while strong induction is problematic. Strong induction yields results suited to analysis, results that involve a completed infinite totality; hence the problem of consistency might arise. "Weak" induction yields results more appropriate to elementary number theory, results that, typically, are not used as the minors of further *modus ponens*.

What about the consistency proof for mathematics, envisioned by Hilbert? Well, it is clear that we could interpret the result, the statement of consistency, using the weak universal quantifier. For Hilbert did not believe that any but a simple induction procedure would be used directly to yield the desired result. And even the intermediate steps of induction need request no strong induction. Intuitively, strong induction is necessary only in analysis, when one deals at once with whole sequences. True, stronger principles of induction are required for the proof than Hilbert thought. But Hilbert may well have been right in his view that the consistency proof for arithmetic requires only weak induction.

It remains only to discuss whether Hilbert used the consistency "result" (that is, his anticipated result) itself in *modus ponens* arguments, as a minor premise, or whether he limited himself to arguments requiring only the weak interpretation of consistency—as his rejoinder to Poincaré (on my interpretation) demands. Let us look at one of the consequences, then, that Hilbert draws from his projected consistency proof. He says:

> But even if one were not satisfied with consistency and had further scruples, he would at least have to acknowledge the significance of the

consistency proof as a general method of obtaining finitary proofs from proofs of general theorems—say of the character of Fermat's theorem—that are carried out by means of the ε-function. . . .

Let us suppose, for example, that we had found, for Fermat's great theorem, a proof in which the logical function ε was used. We could then make a finitary proof out of it in the following way.

Let us assume that numerals $\mathfrak{p}, \mathfrak{a}, \mathfrak{b}, \mathfrak{c}\,(\mathfrak{p} > 2)$ satisfying Fermat's equation $\mathfrak{a}^\mathfrak{p} + \mathfrak{b}^\mathfrak{p} = \mathfrak{c}^\mathfrak{p}$ are given; then we could also obtain this equation as a provable formula by giving the form of a proof to the procedure by which we ascertain that the numerals $\mathfrak{a}^\mathfrak{p} + \mathfrak{b}^\mathfrak{p}$ and $\mathfrak{c}^\mathfrak{p}$ coincide. On the other hand, according to our assumption we would have a proof of the formula

$$(Z(a) \,\&\, Z(b) \,\&\, Z(c) \,\&\, Z(p) \,\&\, (p > 2)) \rightarrow (a^p + b^p \neq c^p),$$

from which

$$\mathfrak{a}^\mathfrak{p} + \mathfrak{b}^\mathfrak{p} \neq \mathfrak{c}^\mathfrak{p}$$

is obtained by substitution and inference. Hence both

$$\mathfrak{a}^\mathfrak{p} + \mathfrak{b}^\mathfrak{p} = \mathfrak{c}^\mathfrak{p}$$

and

$$\mathfrak{a}^\mathfrak{p} + \mathfrak{b}^\mathfrak{p} \neq \mathfrak{c}^\mathfrak{p}$$

would be provable. But, as the consistency proof shows in a finitary way, this cannot be the case. [1928, p. 474]

The reader will see easily that this argument does not make use of the entire range of the consistency statement, but rather proceeds by instantiating in the consistency theorem for a given set of numerals making up a particular formula. It is true that what is proved is a general conclusion: the German letter formula

$$\mathfrak{a}^\mathfrak{p} + \mathfrak{b}^\mathfrak{p} \neq \mathfrak{c}^\mathfrak{p}$$

or, expressed differently, that if $x, y, z, w > 2$ are numbers, then

$$\bar{x}^{\bar{w}} + \bar{y}^{\bar{w}} \neq \bar{z}^{\bar{w}}.$$

If we were to formalize Hilbert's proof in a metalanguage, the overall form of the argument would be:

Mathematical Induction

*(1)	$(x)C(x)$	premise
*(2)	Cx	from (1) by UI
**(3)	$(x)[Cx \supset Dx]$	premise
**(4)	$Cx \supset Dx$	from (3) by UI
**(5)	Dx	from (2) (4) by *modus ponens*
**(6)	$(x)Dx$	from (5) by UG [x flagged]

where $(x)Cx$ is the consistency theorem, and $(x)Dx$ is the result desired, that for every general theorem, there is a finitary proof. The premise $(x)C(x)$ is not used in *modus ponens* until its quantifier is removed.

This example is typical of the way the consistency proof is to be used in mathematical logic, and it reinforces our belief that the consistency proof is not used in ways that force upon us the stronger interpretation of its universal quantifier. But one worry remains: Hilbert may have used the consistency theorem in informal *modus ponens* arguments in *philosophy*, in nonmathematical reasoning in ways which compel the strong interpretation.

For example, in speaking of "ideal elements," Hilbert says: "For there is a condition, a single but absolutely necessary one, to which the use of the method of ideal elements is subject, and that is the *proof of consistency*; for, extension by the addition of ideals is legitimate only if no contradiction is thereby brought about in the old, narrower domain, that is, if the relations that result for the old objects whenever the ideal objects are eliminated are valid in the old domain" (1926, p. 383).

If we assume that the consistency proof is a necessary and *sufficient* condition for the introduction of "ideal elements"—nonfinitary axioms—into the formal system, then the reasoning of Hilbert is as follows:

If the augmented formal system is consistent, then it is justified, i.e., it may be used without qualm.
But the augmented system is consistent.
Hence, it is justified.

It is in informal arguments like these that the consistency of the system appears in the minor premise of *modus ponens*. Even in such cases, however, one must analyze what is meant by the justification of a theory. For it is possible that Hilbert could develop a concept of justification of a theory relative to proofs of a given length. That is, that the system is justified whenever one uses it for proofs of length n or less, n an integer. If there is such a concept, then Hilbert could rephrase the philosophical informal *modus ponens* argument as follows, so that the initial quantifiers might be all interpreted as weak:

If the augmented formal system is consistent, when used in arguments of length n or less, then it is justified for arguments of length n or less.

But, by the consistency proof, the formal system is consistent for arguments of less than length n or of length n.

Stripping off the initial quantifiers, applying *modus ponens* and UG, we conclude:

The formal system is justified for arguments of all lengths.

These last remarks are the outline of a study, nothing more. One must study Hilbert to discover a concept of partial justification for formal theories. One point worth making is that if Hilbert sees ontological gains in consistency proofs—if Hilbert accepts the idea that existence for mathematical objects *is* nothing more than the consistency of the axioms for such objects—then it is hard to see how one could use a concept of partial justification based on the lengths of proofs one is to use. For this would seem to imply that existence admits of degrees. On the other hand, if consistency results have none but epistemic implications, one might more easily accept the concept of partial reliability as a function of the consistency of a system limited to arguments of a certain finite length.

References

Ayer, A. J. 1936. *Language, Truth, and Logic.* 2d ed. New York: Dover, Excerpted and reprinted in *BP.*

Benacerraf, P. 1960. "Logicism, Some Considerations." Ph.D. thesis, Princeton University.

——. 1965. "What Numbers Could Not Be," in *Philosophical Review,* 74: 47–73.

——. 1973. "Mathematical Truth." Unpublished paper. A revised version appeared in *Journal of Philosophy,* 70: 661–679.

Benacerraf, P., and H. Putnam, eds. 1964. *Philosophy of Mathematics: Selected Readings.* Englewood Cliffs, N.J.: Prentice-Hall.

Bernays, P. 1959. "Comments on Ludwig Wittgenstein's *Remarks on the Foundations of Mathematics,*" in *Ratio,* 2: 1–22. Reprinted in *BP.*

Berry, G. 1969. "Logic with Platonism," in *Words and Objections: Essays on the Work of W. V. Quine,* ed. by D. Davidson and J. Hintikka. New York: Humanities Press.

Boolos, G. 1971. "The Iterative Conception of Set," in *Journal of Philosophy,* 68: 215–230.

Carnap, R. 1931. "The Logicist Foundations of Mathematics." Part of a symposium that appeared in *Erkenntnis,* 2: 91–121. (Original title: "Die

References

Logizistische Grundlegung der Mathematik.") Trans. by Erna Putnam and Gerald J. Massey and reprinted in *BP*.

———. 1956. "Empiricism, Semantics, and Ontology," in his *Meaning and Necessity: A Study in Semantics and Modal Logic*. 2d ed. Chicago: University of Chicago Press. Reprinted in *BP*.

Chihara, C. 1966. "Mathematical Discovery and Concept Formation," in *Wittgenstein: The Philosophical Investigations*, ed. by G. Pitcher. Garden City, N.Y.: Doubleday.

Church, A. 1956. *Introduction to Mathematical Logic*. Mathematical Series, vol. 17. Princeton, N.J.: Princeton University Press.

Cohen, P. 1966. *Set Theory and the Continuum Hypothesis*. New York: W. A. Benjamin.

Courant, R., and H. Robbins. 1941. *What Is Mathematics?* New York: Oxford University Press.

Cummins, R., and D. Gottlieb. 1972. "On an Argument for Truth-Functionality," in *American Philosophical Quarterly*, 9: 265–269.

Davidson, D. 1967a. "Causal Relations." *Journal of Philosophy*, 64: 691–703.

———. 1967b. "The Logical Form of Action Sentences," in *The Logic of Decision and Action*, ed. by N. Rescher. Pittsburgh, Pa.: University of Pittsburgh Press.

Gödel, K. 1944. "Russell's Mathematical Logic," in *The Philosophy of Bertrand Russell*, ed. by Paul A. Schilpp. New York: Tudor. Reprinted in *BP*.

———. 1947. "What Is Cantor's Continuum Problem?" in *American Mathematical Monthly*, 54: 515–525. Revised and reprinted in *BP*.

Goodman, N. 1965. *Fact, Fiction, and Forecast*. 2d ed. Indianapolis: Bobbs-Merrill.

Goodman, N., and W. V. Quine. 1947. "Steps toward a Constructive Nominalism," in *Journal of Symbolic Logic*, 12: 105–122.

Grattan-Guinness, I. 1970. *The Development of the Foundation of Mathematical Analysis from Euler to Riemann*. Cambridge, Mass.: MIT Press.

Grice, H. P. 1965. "The Causal Theory of Perception," in *Perceiving, Sensing, Knowing: A Book of Readings from Twentieth Century Sources in the Philosophy of Perception*, ed. by R. Swartz. Garden City, N.Y.: Doubleday.

References

Hempel, C. G. 1945. "On the Nature of Mathematical Truth," in *American Mathematical Monthly*, 52: 543–556. Reprinted in *BP*.

Hilbert, D. 1905. "On the Foundations of Logic and Arithmetic," in *Verhandlungen des Dritten Internationalen Mathematiker-Kongresses in Heidelberg vom 8. bis 13. August 1904*. Teubner: Leibzig. pp. 174–185. (Original title: "Über die Grundlagen der Logik und der Arithmetik.") Trans. by B. Woodward in Van Heijenoort, 1967.

——. 1926. "On the Infinite," in *Mathematische Annalen*, 95: 161–190. (Original title: "Über das Unendliche.") Trans. by S. Bauer-Mengelberg in Van Heijenoort, 1967.

——. 1928. "The Foundations of Mathematics," in *Abhandlungen aus dem mathematischen Seminar der Hamburgischen Universität 6*. Pp. 65–85. (Original title: "Die Grundlagen der Mathematik.")

Hilbert, D., and P. Bernays. *Die Grundlagen der Mathematik*. 2 vols. Berlin: Springer, 1934–1939.

Mendelson, E. 1964. *Introduction to Mathematical Logic*. Princeton, N.J.: Van Nostrand.

Newman, J. R. 1956, "Srinivasa Ramanujan," in *The World of Mathematics*, vol. I, ed. by J. R. Newman. 4 vols. New York: Simon and Schuster, 1956–1960.

Parsons, C. 1965. "Frege's Theory of Number," in *Philosophy in America*, ed. by M. Black. Ithaca, N.Y.: Cornell University Press.

——. 1971. "Ontology and Mathematics," in *Philosophical Review*, 80: 151–176.

Poincaré, H. *Science and Method*. Trans. by F. Maitland. Santa Fe, N. M.: Gannon, n.d.

Pollock, J. L. 1967. "Mathematical Proof," in *American Philosophical Quarterly*, 4: 238–244.

Polya, G. 1954. *Induction and Analogy in Mathematics*. Vol. 1 of *Mathematics and Plausible Reasoning*. Princeton, N.J.: Princeton University Press.

Putnam, H. 1967. "The Thesis That Mathematics Is Logic," in *Bertrand Russell: Philosopher of the Century*, ed. by R. Schoenman. Boston: Little, Brown.

Quine, W. V. O. 1936. "Truth by Convention," in *Philosophical Essays for A. N. Whitehead*, ed. by O. H. Lee. New York: McKay. Reprinted in *BP*.

References

———. 1953a. *From a Logical Point of View: Logico-Philosophical Essays.* Cambridge, Mass.: Harvard University Press. 2d rev. ed. 1961.

———. 1953b. "Two Dogmas of Empiricism," in his *From a Logical Point of View: Logico-Philosophical Essays.* Reprinted in *BP.*

———. 1959. *Methods of Logic.* Rev. ed. New York: Holt.

———. 1960. *Word and Object.* Cambridge, Mass.: MIT Press.

———. 1962. "Carnap and Logical Truth," in *Logic and Language: Studies Dedicated to Professor Rudolf Carnap on the Occasion of His Seventieth Birthday.* New York: Humanities Press. Reprinted in *WP.*

———. 1963. *Set Theory and Its Logic.* Cambridge, Mass.: Harvard University Press. Rev. ed. 1969.

———. 1966a. "On Simple Theories of a Complex World," in *WP.*

———. 1966b. "Ontological Reduction and the World of Numbers," in *WP.*

———. 1966c. *The Ways of Paradox and Other Essays.* New York: Random House.

———. 1969. *Ontological Relativity and Other Essays.* John Dewey Lectures Series, no. 1. New York: Columbia University Press.

———. 1970. *Philosophy of Logic.* Englewood Cliffs, N.J.: Prentice-Hall.

Russell, B. 1908. "Mathematical Logic as Based on the Theory of Types," in *American Journal of Mathematics,* 30: 222–262. Reprinted in Van Heijenoort, 1967.

———. 1919. *Introduction to Mathematical Philosophy.* Rev. ed. New York: Simon & Schuster, 1971.

Sellars, W. 1963. *Science, Perception, and Reality.* New York: Humanities Press.

Shoenfield, J. R. 1967. *Mathematical Logic.* Reading, Mass.: Addison-Wesley.

Tarski, A. 1956. "The Concept of Truth in Formalized Languages," in his *Logic, Semantics, Metamathematics,* trans. by J. H. Woodger. New York: Oxford University Press.

Tharp, L. H. 1971. "Ontological Reduction," in *Journal of Philosophy,* 68: 151–164.

Van Heijenoort, J., ed. 1967. *From Frege to Gödel: A Source Book in Mathematical Logic, 1879–1931.* Source Books in the History of the Sciences Series. Cambridge, Mass.: Harvard University Press.

Vlastos, G. 1969. "Reasons and Causes in the *Phaedo,*" in *Philosophical Review,* 78: 291–325.

References

Wigner, E. 1967. "The Unreasonable Effectiveness of Mathematics on the Natural Sciences," in *Symmetries and Reflections: Scientific Essays,* ed. by M. J. Scriven and W. J. Moore. Bloomington: Indiana University Press.

Wittgenstein, L. 1953. *Philosophical Investigations.* Trans. by G. E. M. Anscombe. Oxford: Basil Blackwell.

——. 1956. *Remarks on the Foundations of Mathematics.* Ed. by G. E. M. Anscombe, R. Rhees, and G. H. Von Wright; trans. by G. E. M. Anscombe. Oxford: Basil Blackwell.

Index

Index

Index

Index